Oxford University Press
Walton Street, Oxford OX2 6DP, United Kingdom

OXFORD   NEW YORK   TORONTO
DELHI   BOMBAY   CALCUTTA   MADRAS   KARACHI
PETALING JAYA   SINGAPORE   HONG KONG   TOKYO
NAIROBI   DAR ES SALAAM   CAPE TOWN
MELBOURNE   AUCKLAND

AND ASSOCIATED COMPANIES IN
BERLIN IBADAN

ISBN 0 19 570606 4

First edition 1989

Second edition 1990

Second edition, second impression 1991

Second edition, third impression 1993

OXFORD is a trademark of Oxford University Press

Cover
Detail from a painting by Paul Grendon entitled *Ons vir jou Suid-Afrika.*
This work is being undertaken towards a M.F.A. degree at the Michaelis
School of Fine Art, University of Cape Town.

Published by Oxford University Press Southern Africa,
Harrington House, Barrack Street, Cape Town, 8001, South Africa
DTP conversion by BellSet

Printed and bound by Clyson Printers, Maitland, Cape.

APARTHEID
to
NATION-BUILDING

Hermann Giliomee
Lawrence Schlemmer

CONTEMPORARY SOUTH AFRICAN DEBATES

OXFORD UNIVERSITY PRESS
CAPE TOWN

CONTENTS

INTRODUCTION

This is a book about apartheid: its rise, functioning, and the adaptations to it, and — in the second part — its possible dissolution in the light of the political realities of South African society. As authors, we must admit to a certain hesitancy before embarking on this project. One of the fiercest opponents of apartheid, Dr Neville Alexander, aptly remarked that the literature on apartheid is so extensive today that no single person could study all of it in the span of a lifetime.[1] So why yet another book on a policy which has been hailed by its supporters but more generally pilloried by the media and a growing body of objections, both locally and internationally?

We believe that there are powerful academic and political reasons for another book on apartheid. To begin with, the most pronounced feature of apartheid has been its resilience and chameleon-like quality. Time and again, scholars have made almost dogmatic statements about the goal, nature and rationale of apartheid and its dependence on certain economic and ideological support bases, only to be proven wrong as apartheid has adapted to changing realities.

In 1970, for instance, Frederick Johnstone stated categorically that the 'true rationale of apartheid is to maximise economic development both for the sake of white prosperity and for the material protection of white supremacy'.[2] It was a view which excluded the possibility of the legalization of black trade unions and the formal acceptance of non-racial wage bargaining, or any rapid narrowing of the huge gap in the spending on white and black education. Yet this is precisely what did occur in the two decades after this article was written. It turned out that apartheid was not necessarily identified with maximal white, or white capitalist, material advantage.

With the benefit of hindsight it has become clear that the demand for white prosperity and material protection may not be the sole, or even main, reason for the white support of the policy of apartheid. Whites continued to endorse apartheid in the 1987 election despite nearly three years of severe economic malaise and widespread recognition among the electorate that this situation was substantially due to apartheid. Apartheid and white material privilege coincided earlier in South Africa's history, but the current phase in our political economy suggests powerful contradictions between apartheid and at least three important conditions for white prosperity, namely, internal stability, external credit-worthiness and investor confidence.

The defence of the material interests of the ruling class obviously remains central to South Africa's policy, and will continue to interact with other political motivations, but arguments that it is part of the essence of apartheid are now difficult to sustain. A recent opposition slogan, which stated that a vote for the National Party was a vote for 'national poverty' sums up one telling current political argument.

Scholars have also been compelled to take a new look at the view that apartheid is inescapably linked to a certain Afrikaner mentality. In 1975, Dunbar Moodie[3] argued that the Afrikaner 'civil religion' — a very specific Calvinist belief system and world view — permeated the Afrikaner communal institutions and induced a sense of mission which required a rigidly-defined race policy. Over the past decade, however, this particular 'civil religion' has become increasingly marginal to the National Party (NP) hierarchy. In the 1989 election, as in the two elections prior to this, its pure expression was articulated only by the Conservative Party (CP) and groups even further to the right.

This is not to say that Afrikaner communal identity and needs have been dislodged — indeed, we argue that these remain extremely important. It is, however, manifestly incorrect to assume, as many liberal writers have in the past, that if Afrikaners became 'modernized' and 'secularized', apartheid would wither away. We argue that there is a base of white communal motivation, relevant to the South African conflict and its resolution, which is not peculiarly Calvinist, nor parochial, nor rooted in a Voortrekker past. These elements, which were powerful adjuncts to apartheid in the past, have since become cultural residues.

Stripped of its capitalist and Calvinistic cladding, the essence of apartheid can be studied far more easily. The same is true of the basic commitments of the white group. If, as we argue in this book, apartheid is merely an instrument of Afrikaner and broader white nationalism, what are the real driving forces of this nationalism, and under which conditions, if any, would the Nationalist government be prepared to negotiate a transfer of power to a post-apartheid regime?

This brings us to the second reason for the book, which is political. It would be safe to say that most whites desire the abolition of apartheid, provided it can occur without major upheaval. Early in July 1989, Law and Order Minister, Adriaan Vlok, admitted that South Africa should have rid itself of apartheid long ago, because it had become an albatross around the country's neck. He went on to say that if the government did not kill apartheid and find something to replace it, it would 'crucify us all'.[4]

But what is apartheid? As long ago as the late 1970s apartheid was declared by a cabinet minister, Dr Piet Koornhof, to be dead or dying.

Clearly, by apartheid he meant only the segregated labour order, an all-white Parliament and the most obvious and repulsive features of apartheid — often referred to as petty apartheid — such as separate park benches and the ban on multi-racial sport and theatrical performances.

At the other end of the spectrum we have the African National Congress (ANC) which stated in 1987 that apartheid was nothing other than 'white minority domination through exclusive concentration of political power on the side of the white minority'.[5] This statement is as misleading as that of Koornhof's. In terms of this definition the entire history of white settlement in South Africa, starting in 1652, must be seen as part of the apartheid epoch, which is obviously historically incorrect. Indeed, using the ANC definition, Brazil could also be labelled an apartheid state.

What we have here is not merely a sloppy use of words but a manifestation of the political struggle for control. Benjamin Nimer made the following important comment:

> Underlying the moves and countermoves of the current communal conflict in South Africa is a struggle to control the key terms of discourse — race, nation, apartheid and socialism — by which the conflict is characterized within and outside the country. Although no definition, therefore, of any of these terms can escape politicization, there is a case, historically and with a view to a negotiated settlement, for having apartheid, the most emotion-laden of these terms, limited to post-1948 doctrine and practice. The foundation for a negotiated settlement unaccompanied by overt civil war must be rather detailed agreement, tacit or explicit, on what the end of apartheid means.[6]

The pre-1948 order of segregation and the apartheid system were in some ways quite distinct and in others intimately interwoven. This book attempts to describe apartheid in its historically-exact sense. The first chapter sketches the conditions and policies out of which apartheid was shaped after 1948. Chapters 2 and 3 describe the ideology and functioning of apartheid, and this is followed by a discussion of the adaptations to and partial abolition of apartheid. We call this government intervention 'reform-apartheid' for we believe that South Africa is still well short of a post-apartheid order.

In the second part of the book we look at ways in which South Africa could move towards a post-apartheid era. We focus attention on a framework for negotiation and a possible resolution which could lead South Africa beyond the stage of hegemony by any one particular group or political tradition and towards a united South African nation.

In this regard we make no apology for paying considerable attention to NP policy and views. If there is to be a peaceful settlement, the NP (or its successor) will play a major part in apartheid's dissolution.

We would like to digress somewhat here and briefly discuss the NP's Five

Year Plan which was announced after our final chapter was written, but which has an important bearing upon the views expressed in Chapters 7 and 8.

The link between the central aspect of apartheid and the Five Year Plan must be made clear. Fundamentally apartheid is a political system in which all representation takes place through state-designated racial groups. In its classic phase, apartheid attributed to each of these racial groups a separate sense of nationhood and a political mission to maintain this identity in separate group areas or homelands. At the same time, it set up a system of divide-and-rule which fostered and concealed white domination.

What is significant about the NP's Five Year Plan is that, while it is almost entirely structured around the 'group concept', the motivation for the group as the corner-stone of the policy is given very little substance or rationale. There is no talk of heritage, history or solidarity. The group is not apparently associated with class, status, honour or material achievement, yet it is nevertheless accepted as a taken-for-granted reality and the corner-stone of the programme. The legitimacy of the group concept is based on nothing but itself — and political power.

The NP document proposes a revoking of the formal legal basis for group definition — the Population Registration Act — but asserts the ultimate necessity for, and viability of, the group based on a negotiated definition. It fully acknowledges the inappropriateness of rigid, involuntary group membership and proposes freedom of association and dissociation, while suggesting confidence in the persistence of the major racial categories. The document appeals for unifying values and commits the party to nation-building, without questioning the assumption that group identity will cohere in the process. The authors of the document appear almost over-confident in proposing an 'open' group. The underlying expectation is that the white communal group will remain separate of its own volition.

The suggestions made in the document regarding communal residence patterns or Group Areas, if read in the light of the practical limits to bureaucratic intervention, the housing shortage in black areas, and a fairly massive process of informal integration already underway, clearly spell out a transition away from racial homogeneity in residential areas, for all but those neighbourhoods determined to stay white. Thus, although 'groups' will increasingly lose their present geographical definition, it is assumed that, for political purposes, the white communal group will crystallize out and remain self-consciously mobilized.

One vital prop is promised in the document, namely white political coherence and power. Here the document gives unconditional reassurances: no numerical political domination of a minority group and 'self-determination regarding own affairs'. The option of, say, coloured people *en masse* having themselves reclassified as white is tacitly ruled out. Thus, for all the peeling away of the surrounding or previously coincidental social and economic features of apartheid, the principle of the white communal group as a constitutional category requiring a defined power base is asserted as strongly as ever.

This categorical communal reaffirmation by the major vehicle for white political articulation, the NP, occurs at a time when an ever-widening circle of popular opposition is articulating the ideal of non-racialism more and more forcefully. This concept now also automatically peppers the communications of private sector organizations, even those of a more conservative nature. In other quarters it has spread beyond a circle of convinced liberals and progressives. Just as the NP takes racial 'groups' for granted, so a phalanx of organizations across a wide political spectrum assumes 'grouplessness' to be the only proper basis for political organization. Many of these organizations have turned non-racial grouplessness into a non-negotiable demand for a future South Africa.

Behind the NP's concept of group stands the question of whether or not it can be dissociated from white advantage and institutional inequality. Equally, however, behind the concept of non-racialism lies the question of whether this compelling and attractive ideal is not simply designed to unlock the door for black domination which will only be marginally less exclusive than its predecessor in staffing the strategic positions of power.

While the NP argues that whites need a constitutional prop, the simple removal of that prop in a system of universal franchise creates a framework within which many political forms can emerge. The potentials of majoritarianism include, on the one hand, a genuine power-sharing democracy with counter-balancing interests, and on the other, permanent black hegemony based on African populism, African nationalism or the working masses. The situation could thus be described, essentially, as minority versus majority needs and interests or, to put it differently, the tradition of the group versus non-racial grouplessness.

This new political scene makes a review of apartheid and its major countervailing forces, the mass democratic and Africanist ideals, essential in a context in which, for the first time, the creation of a South African nation has become the major challenge. To our knowledge there is no recently published book which gives full credence to both the minority and majority-oriented traditions in seeking paths away from South Africa's impasse.[7] This work is intended to fill that gap.

While still resilient and potentially lethal, the basic white communal thrust is clearly no longer able to continue its erstwhile institutional and political hegemony. At a recent seminar, Dr Van Zyl Slabbert, one of South Africa's most prominent political analysts and a former leader of the official opposition, predicted that negotiations between the government and the main representative of the mass democratic movement, the ANC, could emerge at any time, and sooner than most people expected. He may well be right.

Yet the success of negotiation requires some clarity of vision about the benefits of an outcome. Negotiation simply forced by circumstances, by political manipulation of opponents, or by well-meaning but coercive external facilitation could lead to drawn-out spoiling tactics, confusion among supporters, strategies to divide opponents and a host of other impediments to a constructive outcome. We have seen these features in over ten years of stop-start negotiations on the future of Namibia.

In the context of negotiations this book has a specific purpose. In suggesting a process within which race-based policy can be transcended, through negotiation, and the challenge of nation-building be addressed, we were strongly influenced by our conviction that a viable resolution to the South African conflict has to take account of the substantive nature of apartheid and the major alternative visions for South Africa. All too often, proposals for change are based on what we would term conceptual negation of the more resilient aspects of the group concept. In this sense, many past analyses have been intellectually confrontational and unstrategic in their declaration that *any* group mobilization on the basis of race or ethnicity smacks of apartheid and is therefore illegitimate.

We would assert that contributions to the resolution of South Africa's conflict can only be effective if they either involve irresistible coercive pressure on the state or, if that is impossible, they take due account of the elements in the state agenda which are so deeply-rooted as to make moral or logical confrontation pointless as a sole strategy. The coercive pressures on the South African state have been powerful in recent times — a major debt crisis, widespread popular rebellion, economic decay and mounting sanctions — and yet the state agenda has not only weathered the pressure but become more sophisticated in its crisis-management as a consequence. Hence we offer the view that an attempt at a correct analysis of the problem is the more strategic approach.

We could not focus our analysis on resolution without some concrete illustrations of possibilities. Our proposals in Chapters 7 and 8 are intended as examples of an approach to resolution rather than as prescriptions. We do not wish to argue for a particular 'solution' as other published contributions have done. These works, offering structured alternatives, have been valuable and have stimulated a great deal of meaningful debate. This volume does not attempt to compete with these. We would prefer our contribution to be seen as an illustration of a *mode* of resolution, and accept that its detailed exemplification can and will be widely varied in the possible execution.

No book on South Africa can escape a note on terminology. We use the term 'black' in an inclusive sense, to encompass the predominantly Bantu-speaking African majority, as well as the coloured and Indian communities.

Finally, we wish to extend thanks both to those who supported the writing of this book and those who tolerated us while it was being written. Our respective spouses, Annette and Monica, showed mild interest and remarkable forbearance. Our children, Francine, Adrienne and Julian, sensibly displayed polite but profound indifference. They clearly want to work towards their own apartheid-free South Africa without their fathers' suggestions. Our canine companions, Jasper and Flippie, Wouter and Wanda, deeply regretted this book for the missed opportunities for walking and playing. Paula Curry was a model typist in finding her way through several drafts.

Hermann Giliomee wishes to thank the University of Cape Town for funding his research project on the nature of the conflict in South Africa, and the Truman Institute for the Advancement of Peace at the Hebrew University of Jerusalem where he spent three months, on a Lester Martin Fellowship, working on this book. We alone are responsible for the views expressed in the book, and although we wish that this could be the last word on apartheid, we would be extremely surprised if this were to be the case.

Hermann Giliomee
July 1989   *Lawrence Schlemmer*

Notes

1   Neville Alexander, *Sow the Wind: Contemporary Speeches*. Johannesburg: Skotaville, 1985, p. 45.
2   Frederick Johnston, 'White prosperity and white supremacy in South Africa today', originally published in *African Affairs* and reprinted in A. Paul Hare et al. (eds), *South Africa: Sociological Analyses*. Cape Town: Oxford University Press, 1979, p. 358.
3   Dunbar Moodie, *The Rise of Afrikanerdom*. Berkeley: University of California Press, 1975.
4   *Cape Times*, 6 July 1989, p. 6.
5   British Parliamentary Human Rights Group, *South Africa and Human Rights*. London: House of Commons, 1989, p. 68.
6   Benjamin Nimer, 'National liberation and the conflicting terms of discourse in South Africa: an interpretation' in *Political Communications and Persuasions*, 3, 4, 1986, p. 313.
7   The first major study of the national question in South Africa which deals with the various theories and strategies is by No Sizwe (pseudonym for Neville Alexander), *One Azania, One Nation: The National Question in South Africa*. London: Zed Books, 1979. This book is still banned in South Africa.

PART ONE

Apartheid and its reforms

CHAPTER 1

Background to apartheid

The apartheid system was not a unique set of ideas and practices that sprang full-blown from the heads of Afrikaner nationalists. It built upon the segregation order which had developed along with the accelerated industrialization of South Africa following the discovery of diamonds (1869) and gold (1886). Race and class relations in the industrial period grew up around the forms of domination and privilege which had arisen from the time of the founding of the settlement in the Cape in 1652.

Early forms of white supremacy

Three main features characterized social relations in pre-industrial South Africa. Firstly, slavery and frontier conquest placed an unskilled and rightless black labour force under white control. No white labour force developed in the pre-industrial period. To do manual labour in the service of someone else meant a disastrous loss of status for a white. Even the poorest of whites considered themselves superior to blacks.

Secondly, there was always a sufficient number of whites to man all the strategic positions in the political, economic and administrative systems of the country. Whites owned nearly all of the land, and staffed the top and medium-level positions in the state apparatus, army and police. Unlike Brazil, South Africa never needed to create a relatively free and more privileged mulatto class to occupy the intermediate positions in the economy which whites did not want and for which slaves were considered unfit.

Thirdly, growing racial discrimination characterized colonial society from the late eighteenth century onwards. Previously, the individual status of the great majority of people was unequivocally determined by their legal status. People in the legal categories of Company servants and free burghers, nearly all of whom were whites, invariably considered themselves to be superior to slaves, free blacks and to so-called aliens (Khoikhoi) all of whom were black. By the turn of the eighteenth century, race had become the most important factor determining divisions within the society and it created new social tensions.

The main conflict areas were those of marriage, the church and public office. Negative comments about mixed couples became frequent and such couples were often ostracized. In 1787 a separate military unit was formed for 'bastards and mistiches', that is, people of mixed blood. By 1788 burghers were refusing to serve under an officer who was of 'blackish colour and of heathen descent'. Discrimination in the church became common. In the second half of the eighteenth century Dutch Reformed Church (DRC) congregations began to keep separate baptismal records for bastards and Khoikhoi, and for slaves. In the 1820s and 1830s rural western Cape congregations started to ask permission to hold separate church services for blacks. This was followed by requests from congregations on the outskirts of the colony for permission to hold services for blacks in separate buildings. In 1857 the DRC synod acceded to this request — 'on account of the weakness of some' as it phrased the motivation — and in 1881 it founded a separate church for coloured people, the Dutch Reformed Missionary Church.

Finally, from the 1780s the Dutch government restricted the rights of people who were nominally free, such as Khoikhoi, bastards and free blacks. They or their children were indentured as labourers, had to carry passes when travelling, suffered restrictions as to where they could live and had to perform certain compulsory services.[1]

In 1828 the British colonial authorities issued Ordinance 50 which abolished all statutory discrimination and made Khoikhoi and free blacks equal to Europeans before the law.[2] By freeing the Khoikhoi and emancipating the slaves in 1838, the colonial government sought to establish a new order of domination in the place of the old one. They hoped that a class structure would replace the one based on legal status and, informally, on race.

In large areas race, which was such a profound dimension of Cape slavery and frontier conquest, remained the effective basis of the social order. Over time, ex-slaves and Khoisan were relegated to a quite different, and inferior, category from whites. As farm labourers they were severely restricted in their movements by a series of 'masters and servants laws' which imposed the compulsory registration of all contracts and criminal penalties for a worker's breach of contract. A mould was set which neither the colour-blind constitution, which was implemented in the colony in 1854, nor industrialization was able to crack. It was a society in which race and class distinctions steadily reinforced one another. Miscegenation continued to occur, but it was mostly of an extramarital kind. It was invariably the poorer whites who married blacks and they were rarely able to provide their offspring with the means with which to become absorbed into white society.

Cape Town did not quite conform to this pattern. Here mixing between whites and blacks occurred more freely. An important reason for this was the existence of a relatively large free black population who, by 1827, comprised about 25 per cent of the free urban population, excluding troops. Like the mulatto class in Brazil, these free blacks of Cape Town and their descendants blurred and softened the line of division between whites and blacks. 'Passing for white' was possible, especially for fair-skinned girls from coloured families of some standing.

On the eastern and northern frontiers it was initially not particularly clear who belonged where, and the colonists found it difficult to uniformly subordinate people who were not white. As Martin Legassick wrote: 'White frontiersmen expected all their dependants (save their families) to be non-white; they did not expect all non-whites to be their servants'.[3] They could, in other words, also be concubines (but rarely wives) clients, partners in hunting expeditions, military auxiliaries, or even allies.

But this fluidity only existed in the initial phases of conquest when the frontier was still open. After the frontier had closed and white domination had been firmly established, social relations became rigid along racial lines. The same development occurred in the Voortrekker societies which were established in the Transvaal and Orange Free State in the 1830s and 1840s. They were characterized by a fierce resistance among whites to *gelykstelling*, that is, attempts by the government or missionary societies to put blacks on an equal footing with whites before the law, in church or in school. The Afrikaner frontier colonists, and also the British settlers in the Eastern Cape and Natal, demanded wide-ranging discrimination to underpin the political order.

Frontier communities rejected social mixing, not because of an immutable belief in their biological superiority, but because they thought it dangerous, politically. This sentiment was well expressed in 1860 by some Lydenburg colonists who declared that 'to expand the doctrine that confirmed Christian natives are of equal status with white men ... will only have the result that converts, and those who are not converts, will become yet more arrogant, haughty, and untameable than previously'.[4] The Voortrekker republics of the Transvaal and the Free State entered the industrial era explicitly committed to a rejection of any equality in either the church or the state.

By the 1870s, when South Africa entered the era of accelerated industrialization, whites had long been accustomed to domination and privilege, and to the exclusion of blacks from the state or church and some residential areas. This was segregation *as domination*; whites wanted blacks apart in order to keep them 'under' and inferior. There did exist, however, within the white community, a small, but not uninfluential, segment who opposed the complete subordination of blacks.

In the 1820s and 1830s missionaries and enlightened frontier administrators began to advocate segregation *as a form of trusteeship*. They wanted to protect African land and traditional culture. The missionary John Philip saw this form of segregation as the only policy which could stem the remorseless conquest of Xhosa society on the Eastern Frontier. Arguing for a firm border between settlers and indigenous peoples to prevent further dispossession, he advocated that Africans be allowed to keep their own customs and laws, and remain autonomous. The relationship between Xhosa and colonial society had to be conducted by treaty on the model of international negotiations.

In the Cape Colony these thoughts were elaborated by an Afrikaner frontier administrator, Andries Stockenström. He believed that in the case of the Xhosa, the colonial government was confronted with a well-organized and consolidated society which made interference in its internal affairs inadvisable. The introduction of English law among the Xhosa was not practicable and the chiefs could not be expected to administer an alien legal system. Any attempt to weaken the authority of the chiefs and traditional tribal institutions would be counter-productive. Stockenström argued that it was only by recognizing and promoting the traditional structures of political authority in tribal communities that a basis could be gained for a stable resolution of the frontier conflict.[5] In the end the colonial government in the Cape Colony opted for a policy of military conquest which was completed by 1879. This made it impossible to implement a treaty system which could put Stockenström's ideas into practice.

In the mid-nineteenth century, another frontier administrator in the colony of Natal was wrestling with a similar problem, dealing with an equally well-organized, indigenous entity, the Zulu. He was Sir Theophilus Shepstone, Secretary for Native Affairs from 1848 to 1875, who developed his own line of segregationist thinking. He persuaded the colonial government in Natal to set aside separate reserves, recognize the authority of the chiefs and the validity of customary law, and minimize contact between Africans and whites. In his approach, the reserves had to provide a shelter for a separate African political life and culture. But Shepstone's shelter remained very small and quite inadequate. Zululand was a mere patchwork of numerous 'locations' and small mission stations where Africans held land.

Segregation as a form of trusteeship never made much headway in the Transvaal or Orange Free State. In the Free State only a few small locations were carved out, and the Transvaal lacked both the will and the resources to establish consolidated areas for larger tribal units. By the middle of the twentieth century apartheid ideologues would enthusiastically propagate segregation as trusteeship as being traditional Afrikaner policy. However, little or none of this endorsement was evident by the turn of the nineteenth century. Afrikaner leaders such as Jan Smuts, Louis Botha and F. W. Reitz wanted the locations to be broken up in order to flush out African labour, especially to the farms. They also called for the eradication of African tribal institutions and culture because these were thought to impede the flow of labour.

In the Cape Colony segregation as trusteeship did not take root immediately in the second half of the nineteenth century, despite the existence of relatively large African territories such as Transkei, Tembuland, Griqualand East and Pondoland. Government policy was imbued with the Victorian 'civilizing mission'[6] in the service of the policy goal of amalgamation. The policy, as formulated by Sir George Grey in 1854, was aimed at introducing among the Xhosa 'institutions of a civil character suited to their present condition, and by these and other means to attempt gradually to win them to civilization and Christianity, and thus to change by degrees our present unconquered and apparently unconquerable foes, into friends who may have a common interest with ourselves'.[7]

Grey's successors also pursued policies of 'amalgamation' and 'identity'. They rejected the conservative doctrine put forward by Shepstone in Natal as keeping Africans in a 'static' and 'barbaric' state. The administrators at the Cape sought to undermine chieftainship and other forms of traditional authority, root out polygamy and *lobola* as pagan customs, and integrate the African people into colonial society. They stressed the great significance of producing a Westernized African élite who could be used in the colonial administration, and a class of individualistic peasant farmers who could be separated from the masses. The entire emphasis of the policy was on individual incorporation.

By the 1880s, however, the initial optimism about the amalgamation policy had waned. The Western world was experiencing an upsurge of pseudo-scientific racism. Frontier wars and rebellions reinforced the image of the 'lower races' being locked into a vicious cycle of barbarism and degradation. In the Cape Colony the tenacity with which Africans clung to their customs gave rise to the pessimistic assessment that 'progress' in the direction of Western culture and civilization would occur extremely slowly, while some even hoped that the 'lower races' or 'savages' might disappear as a result of the onslaught of industrialization and civilization.

The Glen Grey Act of 1894 spelled the death of the amalgamationist policy in the eastern districts of the colony. The law stipulated that Africans in the Glen Grey district would be allotted plots of land of eight acres. The effect was that most people did not qualify for the parliamentary vote. As a form of compensation it set up elective councils to deal with local issues. The councils were dominated by the chiefs. The *kholwa*, Christianized Africans who had been the focus of the previous policy, were left out in the cold. A pronounced shift had occurred from amalgamation to segregation, and from the belief that African and European cultures were compatible, to one that they were not. Far from believing that Africans could be won over to Western civilization and Christianity, the architect of the Glen Grey Act, Cecil John Rhodes, wrote that the Africans were still children, just emerging from barbarism.[8]

Early industrialization and white supremacy : 1870 – 1910

The earliest forms of modern segregation had appeared by the end of the nineteenth century. This was in the first place connected to the demands of the capitalist system, in particular the mining and farming sectors, which responded to the accelerated industrialization of South Africa. Equally important were the political fears of white leaders, mostly Afrikaner politicians, but also men such as John X. Merriman, that white supremacy would be fatally undermined by industrial transformation and rapid urbanization. These two driving forces gave rise to the initial attempts to regulate African labour effectively and entrench white political control.

To understand the traumatic effects of the industrial transformation of South Africa it is important to realize that the frontier was a mixed capitalist and subsistence frontier, dominated by the subsistence element. When the frontier began to close in the 1860s, there was no economic abundance and no wealthy class with a uniquely democratic vision, as was the case in North America. In fact there was an acute lack of capital, a paucity of natural resources, growing cleavages in white society between rich and poor, and a weakly developed institutional structure which struggled to provide a sense of national cohesion and failed to promote economic growth through administrative efficiency and political stability.

The industrialization of South Africa opened up new markets for farmers, particularly in Kimberley and Johannesburg. As this occurred an acute labour shortage arose. To tie down Africans as farm labourers the Voortrekker republics prohibited Africans from travelling without official documentation, but this was never effectively implemented. Instead farmers levied hut taxes, 'apprenticed' African children, and often used violence to intimidate their workers. However, the position of the farmers was undermined by the fact that the mines and the railways offered wages that were four times higher than those earned by farm labourers.

By the turn of the nineteenth century Africans increasingly managed to resist coercion. It was only the prosperous farmers who could attract sufficient African labour. African tenants who paid a cash or a labour rent for living on a farm preferred to attach themselves to larger landholders who owned fertile land where they could sow and pasture their cattle. Africans were often able to demand favourable terms in share-cropping arrangements.

The obverse of the large number of Africans living on the farms of the wealthy was the labour shortage experienced by the smaller landholders and the white tenant farmers, called *bywoners*. Obviously African labourers would have had to work harder on the smaller farms; moreover, there was no status attached to working for a poor man. The republican governments were unable to redistribute farm labour more evenly, and by the turn of the century the smaller farmers were complaining vociferously that they were forced to work themselves and to use 'their children as kaffirs in order to hold their own'.[9]

For the poorer class of farmers this turned into a crisis of grave proportions. Landholders tended to value them less than African 'squatters', who were generally regarded as better labourers and producers under share-cropping arrangements. By the 1890s many of the poorer farmers, especially the *bywoners* or tenant farmers, were squeezed off the land and the drift to the cities had begun. In 1899, soon after arriving in Pretoria, the young Jan Smuts warned that if laws were not passed to protect the white farmers, 'their sons would have to become labourers, and the real sons would become a labouring class'.[10]

The rinderpest of the 1890s and the devastation wrought by the British army in the Anglo-Boer War (1899–1902) greatly compounded the agrarian crisis of the two ex-republics. But the war also stimulated an Afrikaner cultural and political awakening. Leaders, such as General Hertzog in the Orange Free State, sought to salvage their people, in particular the poor, from 'going under' in an economic revolution conducted by an alien imperial power.

Hertzog and other leaders repeatedly warned of the dangers of the whites 'going under' unless they regained control of the land and farming enterprises. They began to urge farmers' congresses to denounce squatting by Africans which made it impossible to uplift the mass of the impoverished whites. In 1911 a letter in the Bloemfontein newspaper, *The Friend*, observed that white *bywoners* could not obtain land on the half-share ploughing system, whereas nearly every farm had natives who ploughed under this system, and in some cases owned more stock than the white man. A farmer told a government commission that 'the half-share system is a pernicious system, because it takes away from your neighbour natives who ought to be servants'.[11] Prior to the Natives Land Act (1913) which addressed these grievances by attacking African squatters, the Prime Minister, Louis Botha, was asked at a meeting whether it was right that there were Africans who were living a life of comparative ease while there were hundreds of poor *bywoners*. In one instance, a landholder was nearly beaten up when he remarked that he could get more out of one of the African families on his farm than out of seven *bywoner* families together.[12]

In the Cape Colony the flight of rural poor whites to towns had begun several decades earlier. From the 1860s onwards reports noted that unskilled and lowly-paid whites were settling on the fringes of eastern Cape and midland towns or in the backyards of town houses. Slum areas sprung up where whites and blacks lived interspersed. By the 1890s this development had manifested itself throughout South Africa. Afrikaner urbanization, which was often a chaotic and humiliating process, had begun.

Gravely concerned, white politicians began to grope for a policy that could order urban space and provide social controls. Uppermost in their minds was a fear for the demise of white civilization. An editorial writer warned in 1893 that 'poor whites and filthy blacks living side by side would lead to white degeneration', while Merriman cried out that the degradation of the white population was one of the great questions 'upon which our existence as a race in this land depend'.[13] All over the country political and cultural leaders expressed concern over the 'poor white problem'. A shift had manifested itself in the ruling class' perception of poverty. White poverty was no longer seen as a class issue, but explicitly as a racial issue. Instead of regarding white poverty as being the result of individual moral failing to be ameliorated by charity, it was now considered a threat to white supremacy which had to be addressed by state intervention.[14]

Another concern also surfaced as Africans flocked to the towns in growing numbers. This was the fear that the urbanization of Africans would lead to a loss of control over them and to the decay of their cultures. A pioneer of new urban controls was Theophilus Shepstone, who also introduced rural segregation in Natal. To confront the problem of job seekers in towns, Shepstone drew up regulations which contained the essence of influx control based upon a pass system. In regulations issued in 1873, Africans seeking day labour in Durban and Pietermaritzburg were required to register and wear a badge in order to stay in town. This pass system was soon extended to cover all Africans in the town. The authorities also attempted to establish a barrack location system for single males.[15]

It was in Kimberley, with its diamond mines, that the first fumbling attempts at segregation were elaborated into a more substantial system. The mining industry established compounds for single African men, partly to prevent diamond smuggling, but also to ensure a large, constant supply of migrant labour. In addition, mine-owners attempted to put all Africans, not accommodated in compounds, into locations. A historian describes the reasoning: ' "Localisation" was seen as a policy of containment, not only against a threatened uprising, but also in response to the mine-owners' demands for control over those Africans "living at large", who were "lazy", possible deserters from the pits and who lived in dirt and plenty without work or visible means of support'.[16]

The Witwatersrand mines adopted the compound system to house migrant labourers. They were rarely seen in Johannesburg and returned home when their contracts expired. In 1895, at the request of the Chamber of Mines, the Volksraad of the South African Republic expanded the pass law in order to stop workers from absconding from the mines, but the Transvaal state was too inefficient to implement it effectively. In the twentieth century the railways, the docks and new industries across the country took over the compound system. Migrant labour, influx control, and compounds became the key institutions on which segregation as an urban system was built. The compounds and the extensive use of migrant labour were innovations of the mining industry, but passes and the practice of forcing black workers to keep to their labour contracts were bequeathed to twentieth century South Africa by farmers in search of cheap, docile labour. As Stanley Greenberg formulates it: 'Dominant farmers bottled up their farm labourers during an extended transition to capitalist agriculture and in the process urged their ideas and methods on the whole society'.[17] The doctrine of *baasskap*, or white supremacy, and a fundamental lack of community between white and black were forged by the settled farming community, but the urban sectors of the economy willingly adopted or adapted to this scheme.

But segregation was more than an attempt to control black labour. It also resulted from the desire of whites to create towns and cities as orderly white islands in Africa, uncorrupted by the 'lower' civilization and 'unhygienic' standards of blacks. By 1900 this vision was starting to break down. The rapid growth of cities created conditions of extreme poverty, disorder and demoralization. A Cape official wrote of 'large numbers of men — Europeans, Malays and raw Kaffirs — all sandwiched together, living in a state of the utmost neglect, disease and vice'. From about 1902 to 1904 bubonic plague struck the major cities, creating an atmosphere of crisis. It gave rise to what Swanson called the 'sanitation syndrome', which was to push urban segregation a step further. In Cape Town, Port Elizabeth, Johannesburg, Durban and other smaller cities, the sanitation syndrome provided a pretext for determining the African's place in town and for reasserting white dominance in the country.

Officials in Cape Town eagerly suggested that the problem could be overcome if Africans were hounded into prison-like compounds on the model of Kimberley, which 'so easily managed Africans'. In Johannesburg, African slums were burned down within hours of discovering the existence of bubonic plague and Africans were moved to a separate location at Klipspruit. In Port Elizabeth the government forced thousands of Africans out of the city and tried to move them into a new location called New Brighton. Although the authorities did this ostensibly for sanitary reasons, the real driving force was the desire to segregate Africans in compounds or locations. This desire had existed before the plague. The Prime Minister of the Cape Colony, W. P. Schreiner, described it in frank terms:

> What we wanted was to get them practically in the position of being compounded ... keep the natives out of harm's way; let them do their work, receive their wages, and at the end of their term of service let them go back to the place whence they came — to the native territories, where they should really make their home.[18]

Between 1902 and 1904 all of the colonies and ex-republics passed legislation compelling Africans to live in segregated compounds and locations under municipal control. The idea that the white cities and the African locations should be sealed off from one another became a main tenet of twentieth century segregation ideology. The question was how to finance the locations. In 1908 the authorities in Durban hit upon the idea of a municipal trading monopoly in the sale of 'kaffir beer' (beer made from sorghum) to an almost literally captive native market.[19] Beer revenues became the key financial support of a comprehensive system of control. The Native (Urban Areas) Act of 1923 provided, on a country-wide basis, for the funding of the administration of locations from the sorghum beer monopoly.

For segregation to function as a modern system of control, an efficient state apparatus and a comprehensive race policy were required. The modern South African state was constructed by Lord Alfred Milner in the aftermath of the Boer War. To provide for a common native policy for the four colonies, Milner appointed the South African Native Areas Commission (1903 – 5). Comprised almost exclusively of English-speaking whites, this commission, in its 1905 report, tied together the practices of segregation in such a way as to influence South African racial policy deep into the twentieth century.

The Commission firstly proposed that the land be divided into white and African areas. By the end of the nineteenth century whites had taken possession of almost all the available land; however, a number of African reserves did still exist. Along the east coast there was the Transkei, Tembuland, Griqualand East and Pondoland. Further north along the east coast lay the patchwork area of Zululand, and numerous reserves and mission stations where Africans held land. In the Free State only a few small reserves had been carved out whereas in the Transvaal, confusion reigned. Here numerous 'black spots' and reserves existed in the northern and western parts, the majority of which were *de facto* areas of black occupation. It was these reserves which the Commission wanted to earmark for exclusive African occupation.

At this stage, no plans were formulated for future additional land requirements.

Secondly, the Commission recommended that African town locations be established for Africans near all of the major labour centres. Pass laws would control the influx from the rural areas to these towns. Only workers would be permitted to live in the locations, and idle or surplus Africans were to be returned to the reserves. The Commission believed that most Africans would remain voluntary migrant workers, staying in the white areas for periods of less than a year.[20]

Thirdly, the Commission addressed the question of African education. It acknowledged the value of a literary education but stressed the overall importance of agricultural and industrial education which would be of 'particular advantage to the Native in fitting him for his position in life'.

Finally, the Commission endorsed political segregation, thus aligning itself with the political system which had existed in Natal, the Free State and Transvaal prior to the Boer War. In these regions Africans, Indians and coloured people were effectively excluded from the vote. The Cape Colony was an exception. Its constitution of 1853 was colour-blind and it had laid down low property and educational qualifications for voters. Although legislation had curbed the 'blanket' African vote in the 1880s and 1890s, the number of African voters in the Cape Colony was increasing, holding the balance in several constituencies. The Commission did not like the long-term prospects of this, which, if allowed to continue, meant African majority rule. So, to correct this tide, it proposed separate political systems for whites and Africans. The Commission had little place for the educated African élite. In its view, African tribal institutions represented the authentic African voice; hence it advocated that these institutions should be strengthened by bolstering the position of the chiefs and assigning more functions and powers to them.

Nevertheless the Commission did recognize the need for adaptation. It advocated the introduction of native councils in the reserves, comprising elected representatives and appointed chiefs. African interests in a prospective white parliament had to be represented by a fixed number of white representatives elected on a separate roll. As was the case with Lord Lugard's *Dual Mandate*, the classic treatise of the time on the proper government of 'backward races', the Commission's report was imbued with the spirit of trusteeship, a policy regarded as beneficial to both whites and Africans. Cell captures the similarities between the Commission's report and the *Dual Mandate:*

Both professed to have devised means of hearing the authentic African voice. Both excluded the educated African élite as much as possible. Both were based on the presumption that the African racial genius was innately distinctive. Both claimed to be protecting African tradition against becoming pseudo-European. Both aimed to guide the evolution of Africans on their own lines.[21]

The implementation of this vision, of course, depended on the issue of political power. Prior to the Anglo-Boer War, British politicians had promised that victory over the republics would bring 'equal laws, equal liberty to all'. However, the Treaty of Vereeniging, which ended the war, guaranteed that no change would be made to the political status of Africans until some form of self-government had been granted to the ex-republics. Five years later the two ex-republics were granted self-government without the question of African political rights having been addressed. The chance for reform was irreparably lost. At the National Convention (1908 – 9) the delegates from the Transvaal and Orange Free State, together with those from Natal, strongly resisted the extension of African suffrage to their territories. As a compromise it was agreed that qualified Africans and coloured people in the Cape Province were to be kept on the common roll, but no coloured person or African could be elected to Parliament.

It was a concern for coloured people's rights, much more than those of Africans, which animated the more liberal Cape politicians. They came from a long-held tradition which had refused to discriminate against people on the basis of colour alone. They were also worried about a possible joint black onslaught against white rule. *Onze* Jan Hofmeyr, the Afrikaner leader in the Cape Colony, said in 1909:

> It would be a bad day ... if in addition to protecting our northern borders against the teeming millions of Darkest Africa, we had to be continually on our guard against a malcontent coloured and native population in our midst, outnumbering us by five or six to one. [22]

The liberal whites had led coloured people to believe that if they refused to align themselves with Bantu-speaking Africans they would be spared the humiliation of pass laws and compounds. With coloured people occupying an intermediate position between European and African, a distinct 'coloured' political identity had crystallized by the end of the nineteenth century.[23]

In 1910 the Union of South Africa came into being, joining the four territories into a political system which excluded the great majority of

Africans and only half-heartedly incorporated coloured people. All the ingredients were available for white mischief and political opportunism. Zach de Beer, a liberal politician, remarked in the 1980s that the apartheid order had not come about because the average South African legislator was more wicked than his counterpart elsewhere. It had developed because the constitution under which he had laboured since 1910 had mandated him to respond to white wishes, not black ones.[24] There were blacks who understood their precarious position right from the start. When the Draft Act for the Union of South Africa was published in 1909, the coloured newspaper *A.P.O.* (mouthpiece of the African Political Organization) declared: 'The spirit of the whole Act from beginning to end is indicative of the wilful determination of the framers of the Act to draw a colour line in the unified South Africa'.[25]

The term 'segregation' was not used to describe this policy right from the start. In fact, it was seldom used by the turn of the century, and it never appears in the massive report of the South African Native Affairs Commission. Different segments of white public opinion that urged a colour line (the Natal segregationists, capitalist spokesmen such as Lionel Curtis and Howard Pim, and Afrikaner leaders such as Hertzog and Smuts) advocated different approaches to achieve this end. But by 1910 the groping had ended. The term 'segregation' surfaced over a wide front in the deliberations of the National Convention and on party platforms drafted for the 1910 election. Cell notes: 'An organizing principle had been discovered. An ideology was being created. White supremacy was moving into its highest stage'.[26]

Classic segregation, 1910 – 39

Between 1910 and 1939 the successive Union governments established a comprehensive segregation order which reaffirmed and entrenched the two elements of segregation. On the one hand there was the 'fencing off', as D. F. Malan would later call the attempts to entrench domination along colour lines. On the other hand, and less obvious, there was trusteeship which, especially from the 1930s onward, served as official justification for segregation.

One of the main props of segregation was the Natives Land Act of 1913. This act prohibited Africans (except those in the Cape Colony) from buying or renting any land except in the limited area (7 to 8 per cent of the territory of South Africa) which had been set aside for them in the form of 'reserves'. The government wanted African farm labourers as far as possible to work for cash wages, and rule out 'squatting' or share-cropping. The stranglehold which the farmer established over the worker and his family, through the Land Act, largely immobilized the poorly paid African population on the farms. A few Africans living in 'black spots', that is, small plots of land in the 'white' areas which they had bought prior to the Land Act, temporarily escaped this fate. However, in 1939, legislation was passed which was designed to remove these 'black spots' and relocate all inhabitants in the reserves.

Secondly, segregation was ensured by the various forms of urban control. In the early 1920s the Smuts government had laid down the basic framework for administering the lives of urban Africans. A main spur had been the recommendations of the Stallard Commission (1922) which had advocated a system of influx control for Africans. This was based on the principle that the African was only required in the urban industrial areas 'to minister to the needs of whites' and that he had to 'depart therefrom when he had ceased so to minister'. Stallard's ideas about the impermanence of Africans in 'white' towns had important implications. It heralded the Verwoerd dogma of the 1950s and 1960s which deprived Africans of all political rights in the common (or white) area. It also paved the way for a growing white inclination to see Africans as the mere objects of administration, subject to the arbitrary and discretionary powers of the state.[27]

An important law was the Native Urban Areas Act of 1923. It separated the so-called 'location' from the white town through the establishment of a separate, self-balancing, native revenue account. Furthermore, having established a separate account, the state proceeded to argue that Africans could be excluded from white-funded amenities in white areas for which they had not contributed taxes. African locations developed as dormitory towns, removed in body and spirit from the white towns. As Lacey says, even the word location suggests something more remote and inhuman than suburb or village.[28] It is still a remarkable South African phenomenon that whites are often inclined to forget about the black inhabitants when quoting the population of a town, so effectively have they been insulated from the black community in the township.

Thirdly, there was political segregation. Hertzog, Prime Minister from 1924 to 1939, achieved this objective after a protracted legislative struggle. The Representation of Natives Act of 1936 introduced political segregation throughout the country. It removed African voters in the Cape Province from the common electoral roll, and placed them on a separate roll to elect three white representatives to the House of Assembly. Four senators, elected by electoral colleges, were to represent Africans nationwide. At the same time, the reserves were to be extended to cover 7,25 million morgen, comprising in all 13 per cent of the land area of South Africa.

Fourthly, the segregation order expanded the control over African labour and influx. Previously, historians had interpreted the 1936 law purely in political terms. They saw it as an attempt to establish unquestioned political supremacy over Africans; offering Africans more land in compensation for removing those in the Cape Province from the common roll. However, recent research has shown that in the years prior to this Act, the white ruling class was motivated largely by the need to extend the system of control over Africans in order to supply sufficient labour to the mines and the farms.[29] This was spurred by the deteriorating conditions in the overpopulated reserves, which provided the base for the migrant labour system, and the Depression and drought of the early 1930s which caused great poverty among farmers.

By the 1930s migrant labour had become a pronounced feature of the economy. In the reserves the great majority of males had no alternative but to go out and work. The mining industry operated under the pretext that it could afford to pay the migrant worker only a bachelor's wage and justified this with the fact that the worker's family found subsistence in the reserves. But the reserves had become increasingly incapable of providing a subsistence base for these families and the workers' wages were too low to support them.

The legislation which the Hertzog government passed in the 1920s and 1930s on the one hand tried to improve the resource base of the reserves. This helps to explain why the government agreed to expand the reserves and started talking about land that had to be conserved and of areas that had to be developed. In short, it wanted these reserves to carry more people.

On the other hand the government wanted to curb the flow of Africans to the cities by improving the controls at both the rural and urban ends of the system. At the rural end the problem was that the authority of the old chiefs was rapidly being destroyed along with the decline of the reserves, while the tribes themselves had dispersed or become fragmented. Hertzog set about to bolster the tribal structures. In 1927 the Hertzog government passed the Native Administration Bill which finally committed the state to Natal's system of segregation rather than the Cape's tendency towards assimilation. The Governor-General was made Supreme Chief over all Africans, with authority to appoint native commissioners and headmen, define tribal boundaries, and alter the composition of tribes. Lacey describes the thrust of the policy:

Chiefs or headmen, turned into paid government agents, would be answerable to the state for law and order in their districts; they could be used to enforce proclamations, regulations and controls in their areas; and as tax collectors whose stipend was gauged from the number of taxpayers in their district, out of their own material self-interest they could be relied on to encourage their followers to return home annually. The chief's co-operation was thus vital for getting a forced migrant labour system to work.[30]

In 1929 the state introduced a network of labour districts across the entire country which brought African work-seekers under a system of pass laws and controls. Here were the beginnings of a system which would culminate, decades later, in the Department of Bantu Administration — a veritable state within a state. By the 1930s the system and its controls were still inefficient. This was especially true of the attempt to prevent African farm labour from seeking work in town. In 1932 the government introduced the Native Service Contract Act. It enabled farmers to call on a worker and his family to provide labour services, and introduced penalties to keep workers from escaping these services. In some ways the Act resembled the coercive Hottentot Proclamations before Ordinance 50 of 1828.

Despite this, the African drift from the farms continued. By the late 1930s farmers complained vociferously of a labour shortage and of the fact that the pass laws were not strictly enforced. To thwart farm workers moving to the cities, a 1938 circular prohibited the issue of a pass or work permit to any labour tenant unless it clearly indicated the period during which the tenant was allowed to absent himself from his landlord's service. However, the war years compounded the farm labour shortage. This prompted the farmers to call for stricter controls, but the state suspended pass laws for a period during the war.

Finally, the segregation system attempted to establish a job colour bar. By the end of the nineteenth century it was established custom all over the country, with the exception of the Western Cape, that certain occupations or trades were reserved for whites. Generally, all the poorly paid jobs were performed by Africans — 'kaffir-work' as it was called. Furthermore, employers had learnt from experience on the diamond mines that whites would not work under blacks, could not be consigned to compounds or be disciplined in the same manner as blacks. But the modern South African state was slow to put a job colour bar on the statute book. An exception was the Mines and Works Acts of 1911 and 1926 which specified the number of whites required to supervise African miners, limited the issue of blasting certificates to whites and thereafter reserved thirty-two core occupations for whites. In contrast, between 1906 and 1925, several commissions expressed doubts about the wisdom of excluding blacks from economic opportunities for the benefit of whites.[31]

From the 1920s onwards the position of black workers was steadily undermined. After the white miners' strike of 1922, which the state suppressed with force, the Industrial Conciliation Act legalized trade unions. By excluding 'pass bearers', that is, African men, it gave white unionists a commanding role in determining occupational structure, and access to training and wages.

By the 1930s the main parties in government had come to accept Stallard's view that the growth of a stabilized (that is, family-based) African work-force in the towns had to be stopped. Urban employers, as far as possible, had to employ white workers for skilled and semi-skilled jobs while the low-wage unskilled jobs had to be filled by migrants rather than the settled African labour force. Poor education at any rate made it extremely difficult for Africans to compete in the labour market. In 1943 it was estimated that the state would have had to multiply its expenditure on African education by thirty-six times in order to place it on the same level as white education. Two-thirds of African school children were in their first four years of schooling. As primary education for whites was compulsory, some 20 per cent of the white population attended school against a mere 6 per cent of the African population.[32]

Coloured workers, in particular, suffered as a result of discrimination in education. As early as 1905 the Cape Colony had made primary education compulsory for whites but not for coloureds and, as a result, coloureds found it increasingly difficult to meet the entrance qualifications for trades. In 1922 the Apprenticeship Act laid down minimal educational qualifications for entry into specific trades. The effect was severe. While coloured artisans had dominated over half the skilled trades in the late nineteenth century, this fell to about one-thirteenth of the total by 1961.[33]

In addition, the Pact government (Hertzog's NP in coalition with the Labour Party) which took power in 1924, introduced a 'civilized labour policy'. It was supposed to help coloured workers, but it effectively gave preference to white workers in the employ of the railways, harbours, post offices and elsewhere, setting rates of pay for whites to ensure that they could maintain their living standards. The government also tried to put pressure on the manufacturing sector to employ more 'civilized workers', but private business was not eager to comply.

The segregation order relied on harsh security laws. Resistance to the very harsh conditions under which Africans lived and worked began to gather momentum in the late 1920s. The aim of the Industrial and Commercial Workers Union of Africa (ICU) was to draw Africans into a single union with statutory recognition. At its peak it had approximately 100 000 members, drawn from all over the country. A columnist wrote: 'Rent-racked Natives in the urban locations, underpaid Natives in government employ, badly treated Natives on European farms flocked to join the movement'.[34] At one stage it nearly eclipsed the African National Congress (ANC) which, from 1912, had been championing African rights. After a series of strikes in the late 1920s, the state responded with stiff laws to curb African 'agitation'. It empowered the Minister of Justice to ban meetings and publications and banish people if he feared that their activities would engender feelings of hostility between whites and blacks. In 1930 several ANC leaders were banished and several meetings were banned. For the ANC this was a temporary setback, but it dealt a lethal blow to the ICU which was already on the point of collapse owing to internal crises.

Justifying this whole apparatus of exclusion and discrimination required considerable ideological labour. Within the white political leadership this task fell mostly upon the shoulders of Hertzog and Smuts. Hertzog did not venture much beyond the parameters initially set by the Lagden Commission of 1903 – 5. His biographer claimed that he had made an extensive study of the racial issue, including the racial question in the American South, before drafting the Natives Land Act in 1913. But leading liberals had little doubt that Hertzog's thinking stemmed from the recommendations of the Lagden Commission.[35]

Smuts, on the other hand, performed the task of transforming the original crude ideology of white *baasskap* and portrayed segregation as an acceptable form of trusteeship. In his 1929 Rhodes Memorial Lecture at Oxford, Smuts gave an exposition which reflected the ideological shift. He rejected the view of the African 'as essentially inferior or subhuman, as having no soul, and as being only fit to be a slave'; but he also opposed the converse which meant that the African 'now became a man and a brother'. Having dismissed these radical alternatives Smuts proposed a different solution: 'The new policy is to foster an indigenous native culture or system of cultures, and to cease to force the African into alien European moulds'.[36] Smuts projected the Glen Grey Act, which gave Xhosas in the Transkei a form of local government, as a major progressive step since it broke with older policies which wanted to obliterate African culture and traditional institutions.

In articulating these thoughts Smuts was in line with leading thinkers close to the political establishment. These thinkers tried to develop segregation as a compromise between unqualified repression and assimilation. They included Edgar Brookes, Maurice Evans, C. T. Loram and J. D. Rheinallt Jones. Not all of them consistently propagated segregation, but they all contributed to the shifts in thinking on the race question. On the one hand they rejected racial domination based on a theory of biological racism. By the 1920s and 1930s, most liberal thinkers in South Africa had discarded the doctrines of inherent racial inferiority. On the other hand they knew all too well the white fear of miscegenation, degeneration and swamping. Thus they rejected any assimilation based on classical liberalism which spoke of democratic incorporation on an individual basis. Instead, South African thinkers became fascinated with new anthropological thinking about the concept of culture. They no longer envisaged culture as being determined by race or a static civilization, but as the product of dynamic human development. All development strategies had to recognize the distinctive nature of each culture.[37]

By the 1930s the political leadership, and in particular Smuts, had come to accept the idea that each culture was worthy in its own right and capable of further development along its own separate lines. Accordingly, plural cultures had to be the corner-stone of a policy aimed at the gradual upliftment of the African people. In his analysis of this trend, Dubow comments:

> In drawing upon a wide range of racist assumptions, while at the same time avoiding the strong associations of biological determinism, 'culture' came to function as a more subtle form of 'race'. Its relativism and pluralism was ideally suited to the propagation of policies of differential development of segregation.[38]

This notion of culture linked up with the wider imperial policies of indirect rule and trusteeship. In his 1929 speech, Smuts was at pains to declare that a policy of preserving native cultures and institutions was in accordance with the trusteeship clauses of the Covenant of the League of Nations. He looked forward to the day when the majority of Africans would be in charge of their own affairs, under white supervision. Thus in the place of the old system of direct racial domination he held up a system of indirect rule.[39]

By the late 1930s, however, the English-speaking liberal intelligentsia had moved beyond this line of thinking. They increasingly rejected the idea that trusteeship as the basis of segregation policy offered the prospect of an equitable solution to the racial problem. In a sense the segregationist wheel had turned. Whereas earlier, to be a liberal, a negrophilist and a humanitarian meant to be a segregationist, the reverse was now true.[40] The reason for this change of view was not hard to find. Segregation as trusteeship rarely lived up to its ideal. In practice, African chiefs had to play a subordinate role and, if they refused, they were replaced by more pliable functionaries. Liberals also realized that the industrialization of South Africa, based as it was on a large supply of migrant labour, was steadily impoverishing the reserves. This turned the entire concept of trusteeship into a hollow shell.

A decade of flux : 1939 – 48

A rise in the gold price and, more importantly, the Second World War generated an economic boom which propelled race and class relations in South Africa into a situation of unprecedented flux. Gross industrial output increased from 80 million pounds sterling in 1934 to 400 million pounds sterling in 1947. The booming economy, together with the steady decline of the reserves, sent thousands of rural Africans to the cities in search of better work and wages. As influx control was ineffective, the number of urban Africans nearly doubled between 1936 and 1946, increasing from 900 000 to 1 500 000. To a large extent Africans had become proletarianized. In 1947 Donald Molteno, one of the Native Representatives in Parliament, estimated that the reserves were capable of accommodating about half of the African population on the basis of supplementing income by wage-earning outside; the rest were 'overwhelmingly permanent wage servants of Europeans.'[41] In industry the job colour bar was made largely redundant by the war. In many industries Africans and coloured workers did jobs previously jealously guarded as the preserve of skilled whites. The Board of Trades and Industries expressed praise for an increasingly skilled black work-force.

'Isolation has gone and segregation has fallen on evil days',[42] Smuts brooded in 1942, adding, nevertheless, that people did not realize what a great advantage the notion of trusteeship had been. But within the ranks of the English-speaking intelligentsia, trusteeship was being challenged increasingly by a pragmatic incorporationist liberalism. It was espoused by gifted parliamentarians, most prominently by the Native Representatives — Margaret Ballinger, Edgar Brookes and Donald Molteno, but also by Smuts' heir-apparent, Jan Hofmeyr, by intellectuals such as R. F. A. Hoernlé, and by technocrats who wrote government reports dissecting government policy. Accompanying these 'voices of protest'[43] was a more assertive African leadership in the persons of A. B. Xuma, Z. K. Matthews and S. Ngcobo. Smuts' standing prevented the white liberals from attacking him too harshly, but they did convey to him a growing disillusionment with his party's policies.

These liberals rejected the fundamental aim of trusteeship, which was to preserve an indigenous African culture on the point of collapse. For them the supreme challenge was to integrate all racial groups economically and to build a new nation into which, as Ballinger phrased it, 'every section will fit, with security for its own development and respect for its own capacity to serve the community'.[44]

These liberals were far from radical. Ballinger, when questioned by Smuts as to whether Africans wanted the vote, assured Smuts that he still had possibly a decade in which to prepare for such an eventuality. Hoernlé took a distinctly pessimistic view. He recognized that the existing order of racial domination had to crumble ultimately, yet he saw little prospect of whites allowing growing numbers of educated Africans to vote, enter the professions, join the civil service and enjoy the status of social equals. The most that he hoped for was that government would stimulate a steady process of African advancement through improved education, housing and health services, and expanded rights for African workers in industries. He realized that these things would not, in themselves, shake the pillars of the structure of domination but he thought that they would at least prepare Africans for the new order when it came.[45]

The liberals had hoped to get the Smuts government to make a limited but principled commitment to recognize African rights and claims. In their view, the acceptance of the principle of equal rights for all civilized men would have constituted a real gain. It would have enabled everyone in the country to aspire to citizenship, which could then have developed into 'a broader freedom for all our non-European people'.[46] Secondly, they felt that certain individual rights had to be granted, such as the fundamental right of a person to move freely in search of work, and to live with his family wherever he found work. Thirdly, they believed that common industrial rights and practices had to replace the colour bar; this included trade union rights and the gradual phasing out of race differentiation in the social benefits paid out by the government. In general, liberals stressed that the goal had to be for whites and blacks to learn to co-operate and for government to fit a 'diverse, multi-racial society with good will into [a] frame-work of unity'.[47]

When he said that segregation had fallen on evil days, Smuts meant that Africans could no longer be confined to the reserves in great numbers and that the government would have to allow growing numbers to move to the white areas. Gradually the Smuts cabinet also began to recognize that urban Africans were becoming part and parcel of industrial South Africa.[48] Responding to this the government announced in 1942 that it intended 'to make plans for the Native population, both rural and urban, to participate in the enhanced well-being it desired to secure for the community as a whole'.[49] In terms of this policy the government made provision for Africans to be paid old age pensions, invalidity grants and unemployment insurance (this excluded those working on the mines). The principle was established that Africans had to be included in any social security scheme and in any legislation conferring such benefits. Government spending on the education of Africans was no longer to be determined by the revenue from direct taxes collected from them.

As a result there was a considerable increase in educational expenditure and educational facilities for Africans. In primary schools free education, free books and free school meals were introduced. By the end of the war, provision for Africans in higher education was three times as much as it had been in 1936. The medical profession was opened to Africans, as were many other professions. All these measures indicated a sense of government responsibility towards Africans, and a recognition of their citizenship.[50]

Acceptance of economic integration also compelled Smuts and his party to question the industrial colour bar which blocked African labour. Smuts was particularly worried about the fact that the industrial colour bar enabled white artisans to prevent Africans from building their own houses in the townships while there was a huge housing shortage. In questioning the industrial colour bar and the insecurity of black home ownership in the townships, Smuts was clearly concerned about the impact of statutory discrimination upon the better educated African. 'Don't treat a man according to the lowest level of his colour,'[51] he pleaded.

In the political field the Smuts government cautiously attempted to move beyond old-style segregation. The difference between Smuts and his successors should be made clear. After its victory in 1948 the

National Party (NP) introduced segregated institutions in an attempt to stem the black demand for a common political system; Smuts, on the other hand, wanted to create institutions which could slowly expand African participation in a common system. During a private conversation with Hertzog in 1928, Smuts had proposed a system of qualified franchise, mentioning that qualifications would have to be high enough to exclude the bulk of Africans.[52] Ballinger thought that during the war years Smuts had been moving from a policy of controlling blacks to one of governing them.[53] It is true that Smuts did introduce segregated communal representation for coloured people and Indians, but he did not intend this segregation to become an end in itself, as was the case under the apartheid system.

The black groups and segregation

This section will briefly review the position of the black groups prior to the advent of apartheid. Although they had the vote, by 1948 the coloured people had become a marginal group, poorly integrated into the political system. Numerically too weak to elect independent candidates, they were fêted in elections and then forgotten. Neither the South African Party nor the United Party (UP), which received the greatest share of their vote, allowed them any special role within the party.[54] As a result, a large proportion of coloured voters stayed away from the polls from an early stage, believing that they would benefit little by participating.

In 1930 the government extended the vote to white women and a year later it removed the traditional literacy and property qualifications for white males in the Cape. Coloured voters, as a proportion of the total number of voters in the Cape Province, fell from 20 per cent in the 1929 elections to 6 per cent in 1937, when there were 25 238 coloured voters in comparison with 396 237 whites. Between 1937 and 1945 the number of coloured voters rose to 54 134, the highest figure ever. Electoral officers now began to apply the regulations affecting coloured voters more rigidly, challenging those whose educational qualifications were in doubt. As a result, by the time of the 1948 election the total coloured vote had dropped to 46 051, approximately 10 per cent less than the 1946 figure. More importantly, a pervasive cynicism had set in among these voters.

The United Party rejected forced social segregation of the coloured community but favoured residential segregation which they believed could be achieved through voluntary segregation without 'causing ill-feeling or a sense of grievance'.[55] While fully committed to parliamentary representation for coloured people, the Smuts government tried to address the neglect of the coloured community by introducing a Coloured Affairs Department under the Minister of the Interior. It also set up a Coloured Advisory Council (CAC) to serve as a consultative body. Sections of the coloured community, particularly teachers, strongly opposed this move, fearing that it would pave the road for full segregation. During the war years the white community outside the main cities swung away from voluntary segregation to enforced apartheid. In 1939 the Cape Municipal Congress passed by 133 to 127 votes (mostly Cape Town, Port Elizabeth and Kimberley) a resolution supporting the principle of demarcating separate residential areas for whites and blacks, by compulsion if necessary.

By the 1940s much of the coloured population lived in depressed conditions. In 1946 more than 60 per cent of economically active coloured males were low-paid farm workers or domestic servants, or were unemployed, or in undefined work categories. Of the 14 per cent who were factory workers, 30 per cent were semi-skilled and 55 per cent unskilled. Alcoholism was rife, particularly among farm workers.[56]

In the 'white' areas, coloured people had access to few facilities. They were not served in restaurants or cafés (except with take-aways) and the crowded, segregated canteens of the hotels were often the only places in which coloured men were welcome. For the educated stratum of the group there was virtually nowhere to go. They felt themselves alienated from the poorly paid, often dissolute working classes, and they were rejected by whites. There was little contact between them and their white peer groups. A prominent Stellenbosch academic, who lectured in social work, confessed in 1950 that, although she had grown up in the Western Cape, she had no real knowledge of the coloured community in her town or of their activities in the field of education or in the church.[57] There was virtually no integrated sport.

In the civil service, coloureds were not statutorily excluded, but in practice were employed mainly as unskilled labourers. An exception was the police force which employed coloured, Indian, and African officers as part of the non-commissioned establishment, but these officers exercised authority only over other blacks, and their conditions of recruitment, training and pay-scales were inferior to those of whites. The Cape Provincial Administration employed coloureds as teachers and nurses, but discriminated against them with regard to salaries and pensions. Apart from the teachers and nurses, coloureds were employed only at lower wage levels, mainly as unskilled labourers. The

Cape Town City Council, which included several coloured councillors, ostensibly had the most enlightened recruitment policy and working conditions. Here, too, few coloured employees were employed above the lower grades, and not one in a position of authority over white employees.[58]

In the private sector there was a less overt discrimination against coloured people. There was a widespread feeling amongst coloured people, however, that certain nominally colour-blind measures together with custom discriminated against them. For apprenticeships, an educational standard was set that was far too high for much of the coloured population, whose education, especially at the secondary level, was inferior to that provided for whites. More than half the trade unions refused to accept members who were not white. Coloured skilled workers were paid half or less than half that paid to their white counterparts. There existed virtually no professional stratum in the ranks of the coloured people. According to the 1946 census, there were 151 clergymen, but only thirteen doctors and not one lawyer amongst the coloured population. Leadership was largely in the hands of the teachers.[59]

But there were still significant exceptions to the rule of segregation. Coloured people were free to buy property anywhere in the Cape Province. In Cape Town and some of the larger Western Cape towns a complex set of arrangements prevailed which included some areas free of discrimination. Municipal offices in the Western Cape did not have segregated counters. In the field of recreation, coloured people were admitted to performances in the Cape Town City Hall and a few theatres, although special seating was sometimes provided. Some cinemas in 'white' Cape Town did allow coloured people access, especially those people who could pass for white. Cinemas in the smaller towns in the Western Cape usually had segregated seating arrangements. The South African Library in Cape Town was open to all, as were the art gallery and museum, the municipal gardens and parks. Benches marked 'Europeans only' did not exist.[60]

In 1948, the coloured newspaper, *The Sun*, summarized the situation: 'In the past it has been the accepted thing for the semi-whites to keep to themselves in their places of employment, business and entertainment and wherever else it was expedient for them to keep the false flag flying'.[61]

The small Indian population experienced ever more wide-ranging segregation. The Orange Free State excluded them altogether in the 1880s. In the Transvaal and Natal attempts were made to segregate them in locations, ostensibly for health reasons. In 1885 the Transvaal denied Asiatics a vote and property rights, except in locations. Natal, in the 1890s, effectively excluded them from the parliamentary franchise on the grounds that Indians in India enjoyed no such right.

By 1900 there were only about 10 000 Indians in the Transvaal (against 100 000 in Natal). In 1907 the government passed an ordinance which was aimed at preventing any further influx of Indians into the Transvaal and subjected them to an onerous pass law. Smuts told Gandhi, who led the resistance: 'Your civilization is different from ours. Ours must not be overwhelmed by yours. That is why we have to go in for legislation which must in effect put disabilities upon you.'[62]

In 1914, some of the onerous legislation affecting Indians was amended. Indian shops dotted towns in Natal and the Transvaal, and later also in the Cape Province. Nevertheless, Indians faced continuing opposition from sections of the white population who insisted on their repatriation or, failing that, compulsory segregation. The government was reluctant to act, however, fearing the wrath of India and criticism from other members of the Commonwealth.

At the outbreak of the Second World War pressure began to build in Durban in favour of the imposition of land and trading restrictions on Indians. In 1943 a commission found a significant take-over by Indians in a section of Berea, close to the centre of the city. In 1946 the government restricted Indian ownership and occupation in some areas of Natal. By way of compensation it offered Indians limited communal representation in Parliament. This triggered a storm of protests, from the National Party on the one hand, and from the local Indian community and the Indian government on the other. The Indian government referred the treatment of Indians in South Africa to the first session of the General Assembly of the United Nations.

Nevertheless, Indians were not compelled by law to live in segregated townships. In the post-war years they managed to buy property on a considerable scale in the Cape Province, thus incurring the wrath of the NP. In the Transvaal there were Indian shops in the central business districts of numerous country towns. Pretoria had a fairly wealthy Asiatic bazaar area not far from Church Square. In Johannesburg there was a thriving Indian trading area and residences in Pageview, to the west of the central city.[63] By 1950, 19 per cent of all retail enterprises in the country belonged to Indians, although they formed only 3 per cent of the population. Propaganda against the Indians focused only on a few wealthy merchants. The reality was, however, quite different. As late as the early 1960s two-thirds still had an income below the Poverty Datum Line. But the Indians were in the process of escaping from some of the harshest forms of exploitation. By 1951 only 14 per cent of them were still in agriculture. Although they also faced the colour bar in industry, their level of education had steadily improved which improved their job prospects.

According to the 1951 census, a third of the Africans lived on farms, a third were in the reserves, and a third had settled in towns. Although the conditions imposed by segregation upon coloured people and Indians were harsh, those to which Africans were subjected were considerably worse. Dr F. E. T. Krause, Judge-President of the Orange Free State (1933 – 8), summarized it well: 'The African is all the time a prisoner in the land of his birth, although he might not be confined within prison walls'.[64] To an important extent the fundamentally inferior position of Africans under segregation (and later under apartheid) can be attributed to the Native Land Act of 1913. Although originally intended to provide sufficient cheap labour to farmers, the Act effectively relegated Africans to the position of aliens in their own land.

The position of those on the farms stagnated as a result of the Land Act. By preventing Africans from purchasing or hiring land, or engaging in share-cropping in white areas, it excluded them from the productive part of agriculture. In Sol Plaatje's words, after this Act it became 'a fool's errand for the African peasant farmer to find a new home for his stock and family'.[65] Wages of farm labourers rose very little, if at all, until the 1960s. Living conditions remained depressed and only the bare amenities were provided.[66]

The position of that third of the Africans who lived permanently in the reserves also stagnated or declined. The 1913 Land Act and the 1936 Native Trust and Land Act were not intended to provide Africans in the reserves with a viable alternative. The authorities certainly did not envisage a class of self-sufficient peasant farmers. There was not much chance of individual tenure developing, given the overcrowding and the stronghold of communal tenure. The 1936 Act formally committed the state to the principle of trust tenure instead of individual purchase of land. It made provision for the extension of the reserves by seven and a quarter million morgen in order to avert the crisis of appalling soil erosion, overstocking and disease. However only an additional one and a half million morgen had been acquired by 1940, at which time further purchases were temporarily suspended. In the reserves inhabitants were actually forced to cull cattle and curtail their agricultural activities as part of the effort (well-meaning in part) to improve agriculture.

The Land Act provided the base for the migrant labour system. In the 1936 census, 54 per cent of the male population of the reserves, aged between eighteen and fifty-four, were absent, working as migrant labourers.[67] Ultimately this system had a devastating effect on African welfare. As we have seen, the reserves were never intended to provide full subsistence for their residents. Despite this, the mining industry argued that the wages they paid (plus food and living quarters) need only supply a single man's needs; his family could subsist on a plot in the reserves. Many industries followed this example and preferred to employ migrant labour to stabilized labour because of its relative cheapness and docility. In the early 1930's, the new United Party government (which resulted from a fusion between Hertzog's NP and General Smuts' South African Party) reinforced the migrant labour system by announcing that it would pay for the construction of single quarters only. This prompted local authorities to build compound-like hostels rather than family houses. In 1930 Africans were barred from bringing their wives to town until they had been in bona fide employment for at least two years themselves.[68]

The migrant labour system had a vast distorting effect on African society. On the one hand, it weakened the urban base of Africans. Initially most migrants came to the cities as 'target' workers who sought to attain limited objectives. As a result they did not try to overcome the state's constraints by organizing a strong independent labour movement. At the same time their rural base was being steadily eroded along with the decay of subsistence farming; this came about as a result of overcrowding, the long absence of men and the proximity of a strong white farming sector. The reserves became ever more dependent on migrant remittances to buy not only manufactured goods, but also food produced outside of the reserves. In the end the migrant's ability to bargain for a living wage was destroyed, as was African family life and the prospects for a settled African urban society.

The reserves and the migrant labour system also provided the ideological underpinning for urban segregation. By the 1936 census, Africans had undeniably become part of the 'white' urban landscape. Of the two million people who lived in the nine urban areas in South Africa, 812 786 were Africans. Some 570 000 of these lived in the Witwatersrand area. Nevertheless, the fiction was perpetuated among employers that Africans were people who had their own land and that labourers' families did not have to be provided for in the locations. These locations were bleak places. With a few exceptions Africans were excluded from freehold tenure and thus had little incentive to improve their houses. There were no attractive shops. The government granted trading rights to Africans in the locations but prevented the expansion of businesses in the hope that traders would move to the reserves. Africans generally received such low wages that it was impossible for the locations to pay their own way from rents or taxes. The white-controlled municipalities, in any event, tended to have little interest in the improvement of location life. Inside and outside the locations Africans were subjected to numerous laws which applied only to them. Out of a total of 125 032 Africans convicted in the Witwatersrand area in 1936, only one-fifth was for common misdemeanours. All the other convictions were for technical offences such as contravening the pass laws and possession of native liquor.

In the work-place Africans faced a dual handicap: their poor education and the job colour bar. There had been a considerable increase in the amount spent on African education under the Smuts government, but by 1948, 60 per cent of African children were still not in school.[69] Those who were educated found that they could not undertake skilled work in trades covered by apprenticeship committees as these refused to apprentice Africans. As a result, African boys showed little interest in going to trade schools. African trade unions were excluded from the industrial councils where wage negotiations took place, and the Native Labour Regulation Act of 1911 made strike action by Africans a criminal offence.

Although the 1930s and 1940s saw the advance of a small African élite, the majority of Africans lived under appalling conditions. In 1939 the real wage of an African working on the mines was less than it had been in 1914. In 1946, approximately 65 000 African mineworkers went on strike on the Witwatersrand. There had been a doubling, and even trebling, of work load per shift but the improvement of wages was 'miserable' (Mrs Ballinger's word). The Smit Report on urban Africans reported in 1942 that the average family income was usually half the living wage, and that diet was 'disgracefully deficient'. About a third of the urban workers were unhoused. According to this report, poverty had 'all-pervading ill-effects throughout their social life', while pass laws gave rise to a 'burning sense of grievance and injustice'. Employers and officials were often overbearing, 'refusing to treat Africans as people with dignity and self-respect'.[70] Donald Molteno, one of the Native Representatives, summed up his impressions in 1947:

> Our whole African population has been uprooted. They have been proletarianised, pauperised and demoralised. Those — as yet comparatively few — that have acquired

some measure of education are denied occupational opportunities and effective civil and political rights. Their con-
sequent bitterness bodes ill for the future of the relations of
black and white. Our whole society and economy are being
poisoned by our failure to respond to the challenge that
these conditions present.[71]

The Second World War and the general anticipation of freedom for
Third World peoples had made the African leadership more assertive.
The Natives Representative Council, which was established in 1936,
had become increasingly impatient with its own defects. In 1946 Smuts
remarked in a private letter: 'The Natives want rights not improve-
ments'.[72] He was quite correct. When Hofmeyr addressed the Natives
Representative Council in that year, it rejected his defence of government actions, because it had made no attempt to deal with the pass laws,
the colour bar in industry, and political rights, and did not recognize
the African Mineworkers' Union.[73] The Council demanded direct
representation on all levels, from the municipal councils to Parliament.
For Smuts, effective political power for Africans was an unthinkable
concession. He told Edgar Brookes that 'our native policy would have
to be liberalized at modest pace but public opinion has to be carried
with us.' Until this was secured, his approach was 'practical social
policy away from politics' which would be carried out 'more and more
as finance permits.'[74]

The Afrikaner nationalists and segregation

Apartheid cannot be understood without taking the Afrikaner nationalist concerns about material interests and social status into account.
The National Party (NP) of D. F. Malan, which was founded in 1934
in the wake of the fusion of the parties of Smuts and Hertzog, made the
fate of the poorest section of the whites a special plank in its platform.
By 1939 some 300 000 whites were considered to be living in 'terrible
poverty'. The great majority were Afrikaners, who were just over a
million strong at that stage. By the end of the war the general welfare
of whites had improved owing to the booming economy, but the NP
felt that the position of many of the Afrikaners was still precarious.
Approximately 41 per cent were in blue-collar and other manual occupations while only 27 per cent were in white-collar occupations; the
rest were in agriculture. Demobilization of the troops created a fear of
redundancy. Many black blue-collar workers had acquired skills in the
war years and there was widespread concern in NP ranks that a stronger
and more aggressive business sector would employ blacks rather than

whites at cheaper rates. The times were uncertain. The base of the economy was shifting from mining to manufacturing. Secondary industry was rationalizing and replacing the artisan with semi-skilled machine operators. The Board of Trade and Industries in 1945 advocated increased mechanization in order to derive the full benefit of the large resources of comparatively low-paid black labour.[75]

Referring to Afrikaner urbanization in the early twentieth century, the great historian, De Kiewiet, wrote that 'at the base of white society had gathered, like a sediment, a race of men so abject in their poverty, so wanting in resourcefulness, that they stood dangerously close to the natives themselves.'[76] In 1950, two years after his party's historic electoral victory, D. F. Malan spoke in a similar vein. He referred to the great, national, urban migration of both the Afrikaners and the Africans in the first half of the twentieth century and then added:

> The competition was lethal for the white civilized man in the unskilled labour market if he wanted to maintain his civilized standard of living ... He had no other choice than to economise on his food and clothes, and especially on his house. He had to sink down *(afsak)* with his wife and children and there was only one place — the mixed slums, the last resort.[77]

In the 1948 election, the Nationalist Party, under Malan, secured a narrow victory by capturing a critical number of white working-class seats on the Witwatersrand.

The NP feared that under United Party (UP) policy the dividing lines between the coloured people and whites would become increasingly blurred, thus posing a threat to the Afrikaner claims to power and privilege. The NP found the increase in the number of coloured voters from 25 238 in 1937 to 54 134 in 1945 worrying, especially as the steady improvement in the educational qualifications of the coloured people was making ever greater numbers eligible for the vote. The NP had no assurance that this would not be turned to the advantage of the UP, which traditionally drew the overwhelming majority of coloured votes. In the 1948 election, the NP won by only eight seats and they feared that the UP would register large numbers of coloured voters in order to reverse its defeat. It was a situation fraught with uncertainty, as the following statistic of the 1948 election illustrate: out of sixteen constituencies in which coloured voters formed more than 10 per cent of the electorate the UP retained ten seats, lost five to the NP, and captured one from the NP.[78]

In the economic and social fields, Afrikaner nationalists resented the advance of some coloured people, made possible by the absence of any formal racial classification. As we have seen (pp. 24 – 6) this group suffered severe discrimination, both in a formal and informal way. To many fair-skinned coloured people 'passing for white' was the only way out of this maze of discrimination in which they were trapped. The extent of this is reflected in the complaint of the 1936 census takers that it was impossible to calculate the exact number of coloured people because so many coloured parents were attempting to have their children enumerated as whites.[79] The NP was determined to close this escape hatch which allowed people of colour to merge with whites.

Finally, the apartheid order arose from the NP's concern that white political control and Western civilization in South Africa were being steadily undermined by the rapid industrial integration of Africans. The rapid growth of the industrial sector from the early 1930s posed challenges to white domination which had not existed when the economy still relied heavily on the mining and agricultural sectors. Secondary industry, however, depended on a semi-skilled, settled labour-force which was made up largely of Africans who had little intention of ever returning to the homelands.

The proportion of Africans in the work-force of secondary industry grew steadily from the 1920s. In 1930 there were 90 517 Africans in manufacturing; by 1955 the figure had risen to 342 150. Even more significant were the relative figures for whites and Africans. Between 1925 and 1960 the number of African employees in the manufacturing industry increased by 538 per cent, compared with a 271 per cent increase for whites. In 1928 there were only 1,01 African industrial workers for every white; by 1956 the proportion was 2,2 : 1. The urbanization of African women was equally dramatic. For instance in Johannesburg, in 1910, there were 23 African males for every African female, but by 1927 the proportion was 5 : 1, and by 1960 this had narrowed to 1,09 : 1. In the early 1930s Africans formed a larger group than whites in all of the major cities. By 1946 there were 1 810 500 urbanized Africans in comparison with 1 740 800 urbanized whites.[80]

In 1945 the eminent former editor of the *Cape Times*, B. K. Long, warned whites that no matter what they did for Africans, a long-term solution would have to be reached about the political position of Africans.[81] The NP argued that a permanently settled African population residing in 'white' South Africa, enjoying numerical superiority to whites, could stake a strong moral claim to an African majority government, while on the level of power politics, the steady entry of Africans to the strategic levels of the economy posed a long-term threat to white political control.

The NP was convinced that whites would never submit to a black majority even if it meant a 'fight to the death' or a 'bloodbath'.[82] It believed that apartheid was necessary for the sake of both white survival and racial peace. They were determined to stop African urbanization, extend migrant labour to secondary industry, and deflect African political claims to the homelands. The NP thought that the UP was weak and vacillating on all these issues. Although the Fagan (Native Laws) Commission Report of 1948 appeared too late to influence the election, the NP believed it reflected the extent of reformist influence within the UP. Nationalists even suspected Smuts who had remarked with respect to African urbanization that one 'might as well try to sweep the ocean back with a broom'.[83] Now Fagan proposed the stabilization of African labour in the towns, which meant encouraging workers to settle with their families in town. The report concluded that migrant labour was socially and economically undesirable. Three months after he had lost the 1948 election, Smuts told Parliament that the Fagan Report had put the situation before the country with 'a force and a convincingness' that he had not seen previously in any public document on South Africa. He went on to say:

> The native will come more and more into the white areas. The native is not confined merely to the reserves as his home, but he is part and parcel of industrial South Africa.... The native has been integrated into our industrial system and into our economic system. He is our worker; he works on a lower level. He is not a competitor in that sense of the white man, but he is part and parcel of the whole which constitutes South African economic society.[84]

The NP's own report, which appeared in 1947, spelled out a completely different vision. A study group under the chairmanship of P. O. Sauer recommended that Africans should be allowed in the white area only on a temporary basis and that they should be made to retain their links with the reserves. For this reason influx control had to be tightened and migrant labour extended.

Conclusion

On the eve of the 1948 election, the NP published a pamphlet entitled *National Party's Colour Policy*, which contained the following subtitles:

- Maintenance of white race as highest goal
- Welfare of blacks in developing separately

The pamphlet rejected the policy of *gelykstelling* (equalization) of all 'civilized' and 'developed people' within the same political structure and instead proposed apartheid as the only 'guarantee for racial peace'. It also called for the maintenance of a 'pure white race'.

This policy was emphasized by D. F. Malan in his final statement before the election when he spoke of the European race's 'mastery, its civilization and its purity'. He stated further that the NP proposed guaranteeing 'the character of each racial group', and that all groups would obtain the opportunity 'to develop in their own areas into self-sustaining ethnic units [*selfversorgende volkseenhede*]'.[85]

This combined the two ideological themes of white supremacy as it had unfolded in South Africa. The main thrust was 'segregation as domination', which insisted on white domination and the exclusion of blacks. The second theme was 'segregation as trusteeship', which purported to reject the oppression and exploitation of blacks since they would supposedly be given the opportunity to express themselves within their own areas and communities. Although the policy of apartheid was based on both these themes, it was also different. Drawing on the Afrikaners' own ethnic mobilization, apartheid ideologues projected on to the other groups similar ethnic and nationalist aspirations. Helping other groups to realize these aspirations was a justification for depriving them of political rights in the white political system.

This emphasis on the political dimension has prompted some liberal historians to interpret the advent of apartheid in 1948 as the rise of a fundamentally irrational and anti-capitalist policy. They believe that in reasserting the racial division, apartheid harked back to the crude racism of the frontier, or that in denying economic realities, the nationalists misconceived their own interests. Revisionist historians have strongly attacked these interpretations. They argue that capitalist development and apartheid were compatible, at least for the first quarter of a century after 1948. To revisionists, Afrikaner nationalism and apartheid are ideologies that developed as a result of the particular manner in which Afrikaners tried to accumulate capital.

It is true that apartheid and capitalism did manage to coexist, at least until the early 1970s. But historians should always be careful of confusing effects with intentions. The NP's aim, in the run-up to the 1948 election, was to end the political drift which had characterized a decade of flux. National Party leaders had no doubt as to what they would do if confronted with a real choice between political control and economic growth. Verwoerd, the main architect of apartheid, warned in 1951 that the unhealthy concentration of African labour due to industrial centralization 'could bring about the death of white civilization in South Africa'. He added: 'The survival of white civilization in South Africa is of more importance to me, even more important than expanded industrial development.'[86]

Apartheid had to justify both white domination and wide-ranging political intervention in the economy. It was for this reason that the Nationalists spent so much time on an ideology aimed primarily at drawing the Afrikaners together as a cohesive new ruling group.

Notes

1. For a full analysis of the persistence of the racial structure see 'The origins and entrenchment of European domination at the Cape, 1652 – 1840', in Richard Elphick and Hermann Giliomee (eds), *The Shaping of South African Society 1652 – 1840*. Cape Town: Maskew Miller Longman, 1989.
2. For an excellent new interpretation of the modernization of the Cape's administrative structures see the chapter by Jeffrey Peires, 'The British and the Cape, 1814 – 1834', in *The Shaping of South African Society*.
3. Martin Legassick, 'The Frontier tradition in South African historiography', in Shula Marks and Anthony Atmore (eds), *Economy and Society in Pre-Industrial South Africa*. London : Longman, 1980, pp. 67 – 8.
4. J. D. Huyser, 'Die Naturelle-politiek van die Suid-Afrikaanse Republiek'. D. Litt. dissertation, University of Pretoria, 1936, p. 235.
5. Andre du Toit and Hermann Giliomee, *Afrikaner Political Thought, vol I: 1780 – 1850*. Berkeley : University of California Press, 1983, p.140.
6. Saul Dubow, 'Race, civilisation and culture: the elaboration of segregationist discourse in the inter-war years', in Shula Marks and Stanley Trapido (eds), *The Politics of Race, Class and Nationalism in 20th Century South Africa*. London : Longman, 1987, pp. 71 – 9.
7. D. M. Schreuder, 'The cultural factor in Victorian imperialism: a case study of the British "civilizing mission"', *Journal of Imperial and Commonwealth History,* IV, 3, 1976, pp. 283 – 317.
8. Richard Parry, '"In a sense citizens, but not altogether citizens" : Rhodes, race and the ideology of segregation of the Cape in the late nineteenth century', *Journal of African Studies*, 17, 3, 1983, pp. 377 – 91.
9. B. J. Kruger, *Diskussies en Wetgewing rondom die landelike arbeidsvraagstuk in die Suid-Afrikaanse Republiek met besondere verwysing na die Plakkerswette 1885 – 1899,* Communications of the University of South Africa, C62, Pretoria, 1966, p.10.
10. This section is based on Hermann Giliomee, 'Processes in development of the South African Frontier', in Howard Lamar and Leonard Thompson (eds), *The Frontier in History : North America and Southern Africa Compared*. New Haven : Yale University Press, 1981, pp. 76 – 122. The quote appears on p.110.
11. T. R. H. Davenport, *South Africa : A Modern History.* Johannesburg: Macmillan, 1987, p. 544.
12. Timothy J. Keegan, *Rural Transformations in Industrializing South Africa.* Johannesburg : Ravan Press, 1986, pp. 180 – 2.
13. Colin Bundy, 'Vagabond Hollanders and Runaway Englishmen : White poverty in the Cape before poor whiteism', in William Beinart et al. (eds), *Putting a Plough to the Ground : Accumulation and Dispossession in Rural South*

Africa, 1850 – 1930. Johannesburg : Ravan, 1986, pp. 120 – 2.

14. Bundy, 'Vagabond Hollanders', pp. 101 – 28.
15. Maynard Swanson, 'Urban origins of separate development', *Race*, X, I, pp.31 – 40.
16. Rob Turrell, 'Kimberley : labour and compounds, 1871 – 1888', in Shula Marks and Richard Rathbone (eds), *Industrialization and Social Change in South Africa : African Class Formation, Culture and Consciousness, 1870 – 1930.* London : Longman, 1982, p. 56.
17. Stanley Greenberg, *Race and State in Capitalist Development : Comparative Perspectives.* New Haven : Yale University, 1980, p. 387. This is also the view of George Fredrickson, *White Supremacy : A Comparative Perspective.* New York : Oxford University Press, 1988, pp. 219 – 20. For a different view, holding that segregation had its roots in the consolidation of the gold industry, see John W. Cell, *The Highest Stage of White Supremacy : The Origins of Segregation in South Africa and the American South.* Cambridge : Cambridge University Press, 1982, esp. pp. 12 – 18, 46 – 58.
18. This and the preceding paragraph are based on Maynard W. Swanson, 'The Sanitation Syndrome : Bubonic plague and urban native policy in the Cape Colony, 1900 – 1909', *Journal of African History,* 18, 3, 1971, pp. 387 – 410. The quotes are on pp. 395 and 400.
19. Swanson, 'Urban origins', p. 39.
20. Cell, *Highest Stage,* p. 204. These paragraphs are drawn from Cell's sensitive discussion of the commission.
21. Cell, *Highest Stage,* p. 208.
22. J. H. Hofmeyr, *The Life of Jan Hendrik Hofmeyr (Onze Jan).* Cape Town : Van de Sandt de Villiers, 1913, p. 629.
23. Ian Goldin, *Making Race : The Politics and Economics of Coloured Identity in South Africa.* Cape Town : Maskew Miller Longman, 1987, pp. 3 – 27.
24. Zach de Beer, 'The 1986 Constitutional Crisis', *Sunday Times,* 2 January 1986.
25. R. E. van der Ross, *The Rise and Decline of Apartheid : A Study of Political Movements among the Coloured People of South Africa, 1880 – 1985.* Cape Town : Tafelberg, 1986, p. 43.
26. Cell, *Highest Stage,* p. 213.
27. Alf Stadler, *The Political Economy of Modern South Africa.* Cape Town : David Philip, 1987, p. 89.
28. Marian Lacey, *Working for Boroko : The Origins of a Coercive Labour System in South Africa.* Johannesburg : Ravan, 1981, pp. 252 – 3, 368.
29. Lacey, *Working for Boroko,* pp. 277 – 300.
30. Lacey, *Working for Boroko,* pp. 284 – 5.
31. Merle Lipton, *Capitalism and Apartheid: South Africa, 1910 – 1986.* Aldershot: Wildwood House, 1986, pp.18 – 20 and also Greenberg, *Race and State,* pp. 79 – 85, 176 – 81.
32. Edward Roux, *Time Longer than Rope : A History of the Black Man's Struggle for Freedom in South Africa.* Madison : University of Wisconsin Press, 1964, p. 343.
33. Davenport, *South Africa: History,* p. 533.
34. Lacey, *Working for Boroko,* pp. 246 – 7.
35. John David Shingler, 'Education and political order in South Africa', Ph.D. dissertation, Yale University, 1973, p.18.
36. J. C. Smuts, *Africa and Some World Problems.* Oxford: Oxford University Press, 1930, pp. 77, 84.
37. The two paragraphs preceding the reference number draw on the stimulating analysis of Dubow, 'Race, civilisation and culture', pp. 71 – 94.

38. Dubow, 'Race, civilisation and culture', p. 87.
39. Smuts, *Africa and Some World Problems*, pp. 83, 85.
40. Shingler, 'Education and political order', p.17.
41. Phyllis Lewsen, *Voices of Protest : From Segregation to Apartheid, 1938 – 1948*. Johannesburg : Ad Donker, 1988, p. 255.
42. J. C. Smuts, *The Basis of Trusteeship in African Native Policy*. Cape Town: SA Institute of Race Relations, 1942, p.10.
43. This is the title of a thought-provoking collection of the speeches and writings of liberal spokesmen in Lewsen, *Voices of Protest : From Segregation to Apartheid, 1938 – 1948*.
44. Lewsen, *Voices of Protest*, p. 206.
45. Lewsen, *Voices of Protest*, pp. 97 – 9.
46. Margaret Ballinger in Lewsen, *Voices of Protest*, p. 252.
47. Edgar Brookes in Lewsen, *Voices of Protest*, p. 292.
48. *HAD*, 1948, cols. 3474 – 79; *HAD*, 1948 (second session), cols.199 – 203.
49. *HAD*, 1943, col. 6.
50. Margaret Ballinger, *From Union to Apartheid : A Trek to Isolation*. Cape Town: Juta, 1969, p.128 and also Lewsen, *Voices of Protest*, pp. 48, 276 – 8.
51. Cited by Ballinger, *From Union to Apartheid*, p.132.
52. Davenport, *South Africa: History*, p. 295.
53. Ballinger, *From Union to Apartheid*, pp.119 – 28.
54. Carter, *Politics of Inequality*, p.126.
55. *HAD*, 1939, col. 2322.
56. Patterson, *Colour and Culture*, p. 356.
57. Erika Theron, 'Die Kleurling en die houding van die blanke', *Tydskrif vir Rasse-aangeleenthede*, 1, no. 4, 1950, pp. 21 – 4.
58. Patterson, *Colour and Culture*, pp. 65 – 71.
59. Patterson, *Colour and Culture*, pp. 71 – 85.
60. Patterson, *Colour and Culture*, pp. 123 – 9.
61. *The Sun*, 27 August 1948, editorial.
62. W. K. Hancock, *Smuts: The Sanguine Years, 1870 – 1919*. Cambridge: Cambridge University Press, 1962, p. 346.
63. Davenport, *South Africa: History*, pp. 351, 379 – 80 and also Nigel Mandy, *A City Divided : Johannesburg and Soweto*. Johannesburg : MacMillan, 1984, pp.122 – 3.
64. Lewsen, *Voices of Protest*, p. 120.
65. Cited by Francis Wilson, 'Farming, 1866 – 1966', in Monica Wilson and Leonard Thompson (eds), *Oxford History of South Africa*, vol. II. Oxford : Oxford University Press, 1971, p. 130.
66. Wilson 'Farming, 1866 – 1966', pp. 153 – 71.
67. Lacey, *Working for Boroko*, p. 392.
68. Lacey, *Working for Boroko*, pp. 269 – 70.
69. Lewsen, *Voices of Protest*, p. 288.
70. Lewsen, *Voices of Protest*, pp. 42, 51 – 2, 234.
71. Lewsen, *Voices of Protest*, pp. 266.
72. Lewsen, *Voices of Protest*, pp. 53 – 4.
73. Davenport, *South Africa: History*, pp. 343 – 4.
74. Davenport, *South Africa: History*, pp. 343 – 4.
75. Martin Legassick, 'Legislation, ideology and economy in post-1948 South Africa', *Journal of Southern African Studies*, vol.1, no. 1, 1974, pp. 9 – 10.
76. C. W. de Kiewiet, *A History of South Africa : Social and Economic*. Oxford : Oxford University Press, 1957, pp. 181 – 2.
77. FAK, *Tweede Ekonomiese Volkskongres : Openingsrede deur D. F. Malan*. Bloemfontein, 1950, pp. 11 – 12.

78. Gavin Lewis, *Between the Wire and the Wall : A History of South African 'Coloured' Politics.* Cape Town : David Philip, 1987, p. 258.
79. Lewis, *Between the Wire and the Wall,* pp. 162 – 3.
80. N. J. Rhoodie, *Apartheid and Racial Partnership in Southern Africa.* Pretoria: Academica, 1969, pp. 91 – 108, 154 – 5.
81. Cited by Rhoodie, *Apartheid and Racial Partnership,* p. 89.
82. These were the actual words of D. F. Malan in a 1946 speech: see Lewsen, *Voices of Protest,* p. 314.
83. J. C. Smuts, *The Basis of Trusteeship.* Johannesburg : South African Institute of Race Relations, 1942, p. 10.
84. *HAD,* 1948, cols. 202 – 3.
85. NP Inligtingskomitee, *NP se Kleurbeleid* (c. 1947) and also *The Cape Times,* 24 May 1948, p. 3.
86. Cited by Rhoodie, *Apartheid and Racial Partnership,* p. 98.

CHAPTER 2

The ideology of apartheid

Apartheid has been part of South African life for over forty years. As an ideology, distinct from segregation, it dates back to the 1930s when church leaders, academics, journalists and politicians of D. F. Malan's National Party (NP) started formulating the underlying principles. In 1948 the slogan 'apartheid' helped to sweep the NP into power. In the years following, apartheid became a standard entry in the world's political lexicon. Today it is arguably one of the few political terms known throughout the world.

Given the veritable mountain of literature on the subject of apartheid, its ideology has not yet been exhaustively analysed. The best academic analysis by an outsider who consulted primary Afrikaans sources was written two decades ago.[1] Among South African studies, the most outstanding are the 1959 work of Rhoodie and Venter, and Rhoodie's 1969 study.[2] These studies appeared when the ideology was still being formulated and do not discuss fully the development of the ideology within Afrikaans church circles or the academic community. This chapter provides a brief analysis of the ideology as it developed over the past fifty years.

It must be emphasized at the outset that although apartheid was founded upon the policy of segregation, it was also quite different from it. Whereas segregation implied a horizontal division between races, apartheid envisaged a vertical division between equal ethnic groups or nations. Apartheid also differed from the liberal model which English-speaking opinion-formers of the time used to categorize society. The liberal model is based on the individual who is invested with rights. The apartheid model portrays man as a social being who finds fulfilment only in a community. The individual is seen to have no rights while the *volk* has a God-given right to exist; whatever rights the individual enjoys are derived from the collectivity. Gregor concludes:

> Apartheid as a theoretical system is distinguished from other radical ideologies only when the collectivity with which the individual is identified is specified. Leninism originally identified that collectivity as economic *class*. Fascism identified that collectivity as the *nation-state*. National Socialism identified it as a biological *race*. Apartheid, on the other hand, has identified that community as the historically constituted *volk*.[3]

It is in this sense that apartheid can be considered a distinct development, rather than an extension of segregation.

Apartheid as instrument of Afrikaner nationalism

The NP propagated the ideology of apartheid so vigorously, even fanatically, that many observers in the 1960s and 1970s believed that apartheid had replaced Afrikaner nationalism as the NP's goal and rationale. This was a mistaken impression. It was only in the 1970s, however, that leading apartheid ideologues openly declared that apartheid was not a goal in itself but merely an instrument of Afrikaner nationalism.[4] To put it differently, the political order which the NP constructed after 1948 was aimed at enhancing Afrikaner nationalism by entrenching white political control in South Africa. Through apartheid, Afrikaners governed not only themselves, but also all other groups in the society.

A nationalist movement rarely has clear and coherent ideas about the most desirable political and social order. It is through its association with another, more specific ideology (socialism or fascism, or, in the case of the Afrikaner nationalists, apartheid) that it acquires an action-related system of ideas.[5] The forty years of NP rule have seen the operation of two overlapping belief and ideology systems and it is these which have guided the party in its construction of the political order. There was, on the one hand, the Afrikaner nationalist ideology with its claim to the land and Afrikaner sovereignty in that land. Apartheid, on the other hand, was an operative ideology that spelled out the relations between whites and other ethnic groups (or 'nations') in South Africa in a way that both fostered and concealed Afrikaner rule.

It is impossible to understand the rise and evolution of apartheid without a good grasp of the ideology of Afrikaner nationalism. This ideology urged Afrikaners to set themselves apart from non-Afrikaner whites on the one hand, and the various black groups on the other. We can discern both popular notions and derived notions in this ideology.[6]

The popular notions were comprised of a set of loosely formulated beliefs, values and fears. These included a primary claim to the land, a rejection of *gelykstelling* (the 'equalizing' of white and black) and a fear of being swamped. The 'historic right to the land' was a primal rather than legal claim. It started with the myth of a vacant land at the time of white colonization. This is how the influential commentator, Schalk Pienaar, formulated it in 1960:

> South Africa is by no means territory wrested from its rightful owners by the White man. There were no established Bantu homelands in South Africa when Van Riebeeck landed at the Cape in 1652. The whites moving northwards and the Bantu moving southwards did not meet until more than a century later. If newcomers is the word one wants, then the Bantu are as much newcomers to South Africa as the whites.[7]

Nationalists believed that among the whites the Afrikaners had a special claim to the land and the state. The Afrikaners considered themselves to be the pioneers because they had arrived well before the English-speaking settlers and because the Voortrekkers who set out on the Great Trek opened up the interior. D. F. Malan stressed the idea that the Voortrekkers had laid the foundations of the 'white Christian civilisation', and had bequeathed to posterity, not only a territory, but one which was the bearer of the 'national soul' (*volksiel*).[8]

It was further argued that since the Afrikaners were rooted in South Africa and had nowhere else to go, they had to ensure their survival within the country. The only way in which they could safeguard their culture, language and 'bio-genetic' identity was through exclusive political control over 'white' South Africa. In any unitary state Afrikaners (and other whites) would be swamped. As a cabinet minister expressed it: 'We can pool money, but not our racial characteristics, our traditions or our socio-cultural heritage'. Political integration would mean 'the end of the Western civilisation in South Africa, the political eclipse of whites and everything they have built up over a period of more than 300 years'.[9]

Apart from Afrikaner nationalism's popular notions, there also existed derived notions which included both neo-Calvinist ideas about sovereignty in one's own sphere and the thoughts of Fichte and Herder about the metaphysical qualities of nations. The influence of Calvinism deserves special comment. There is no support for the interpretation, fashionable fifty years ago, that the Afrikaners applied a form of primitive Calvinism to project themselves as a 'chosen people' and that their support for segregation and apartheid was rooted in this. On the other hand, there is continuing debate on the connection between neo-Calvinism and the ideology of Afrikaner nationalism.

Both Hexham[10] and Moodie[11] give a special place to nineteenth century Dutch neo-Calvinism and, in particular, to the Dutch theologian and politician, Dr A. Kuyper, as a formative influence on twentieth century Afrikaner nationalists. Kuyper argued that God created the cosmos as a multitude of circles of life, each circle characterized by its own nature and tasks, free and independent of each other;[12] Kuyper's ideal was that the members of his own neo-Calvinist group should express themselves within their own defined circle. Under the influence of Kuyper, a process of *verzuiling* (pillarization) took place which divided The Netherlands into a nation of separate, largely autonomous groups.

D. F. Malan, who studied in The Netherlands when Kuyper's influence was at its greatest, returned with the conviction that the Afrikaners had to organize themselves separately in all walks of life. Like the Dutch neo-Calvinists he argued that Afrikaner strength lay in separate cultural, religious and political institutions. Some years later Malan was propounding the theory that God had ordained separate nations, each with a unique destiny.[13]

More than fifty years later, by the mid-1970s, Kuyper's thought was still influencing some Afrikaner nationalists. A. P. Treurnicht, in a study entitled *Credo van 'n Afrikaner*, considered the *volk* to be the most important sphere of life, and equated sovereignty in one's own sphere with an autonomous political and cultural life. As an apartheid propagandist, Treurnicht also advocated separate residential areas, political separation and independent homelands for the other 'nations' of South Africa. He believed that this system would grant each nation the chance to fulfil its own vocation; sovereign in its own sphere of life.[14]

In a recent analysis Gerrit Schutte expressed valid doubts as to whether this discourse could truly be called neo-Calvinist Kuyperianism. Kuyper was, after all, concerned with the self-isolation of a religious group on the basis of a specific world-view and distinctive beliefs. Afrikaner nationalists, by contrast, sought the mobilization of Afrikaners on the basis of race and culture. They tried to bring Afrikaners together in ethnic institutions regardless of their religious beliefs and world-views.[15]

It was the metaphysical notions about the centrality of nations in God's creation, rather than neo-Calvinist ideas, that influenced Afrikaner ideologues. Inspired by the thinking of Herder and Fichte, these ideas were introduced into the mainstream of Afrikaner intellectual life by a generation of young Afrikaners who had studied in Europe during the 1920s and 1930s. Most prominent among them were Nico Diederichs (later Minister of Finance), H. F. Verwoerd (later Prime Minister), Piet Meyer (later head of the Afrikaner Broederbond and the South African Broadcasting Corporation) and Geoff Cronjé (later Professor of Sociology at the University of Pretoria). The clearest exposition is found in a pamphlet by Diederichs. His premise is that there is a God-given multiplicity of nations, languages and cultures. An individual's growth to full humanity is only possible if he identifies with his nation, and freedom can only be achieved by realizing oneself in service of the nation. As Diederichs phrased it:

> Mankind is not merely called to be a member of a community; he is also, and especially, called to be a member of a *nation*. Without the uplifting, ennobling and enriching influence of this highest encompassing unity which we call a nation, mankind cannot reach the fullest heights of his human existence. Only through his devotion to, his love for and his service to the nation can man come to the versatile and harmonious development of his full personality. Only in the nation as the most total, most inclusive human community can man realize himself to the full. The nation is the fulfilment of the individual life.[16]

It was these metaphysical beliefs about nationhood, rather than neo-Calvinist thoughts about sovereignty in one's own sphere, that found greatest expression in the ideology of an exclusive nationalism. These nationalist doctrines also provided the point of departure for the apartheid ideology. This united the Afrikaner racist, concerned only with white supremacy, and the Afrikaner idealist, who dreamed about separate but equal nations in South Africa.

Apartheid as an action-related system

Apartheid was a fully-fledged ideology containing goals and principles for the reconstruction of society. The main tenets were the following:

Tenet one:

The basic unit of society is not the individual, but the volk with its God-given right to exist.

Afrikaner nationalists defined *volk* as a collectivity whose members were of similar descent and racial stock, and who shared a common history, culture and sense of destiny. In advancing the centrality of the *volk* and its right to survive, the NP used both secular and religious arguments. Accordingly, each *volk* in the world had the right to maintain itself and develop separately. A leading NP politician declared in 1960:

> Apartheid means that the whites locally desire nothing less and nothing more than what the inhabitants of any state demand, namely that their homeland's claim to constitutional self-determination be guaranteed at all times.[17]

Another strand in the secular argument was explicitly racist. Most prominent among the academics who advanced this argument was Geoff Cronjé.[18] He emphasized the biological differences between races and warned against the disastrous effects of miscegenation. Cronjé used the coloured population as demonstration of the negative consequences of racial mixing. According to him, they were a group who lived in constant friction. This was because the inherent differences between whites and blacks were too large to bridge and because, as 'bastards', they represented a sinful union of elements which could not really become one.[19]

Apart from politicians and academics, newspapers such as *Die Burger* and *Die Transvaler*, the Afrikaner Broederbond and the Afrikaans churches played a major role in persuading Afrikaners that apartheid was a just and sound policy. In 1933 the Afrikaner Broederbond, a secret organization comprising clergymen, academics, farmers, professionals and politicians, gave one of the earliest formulations of the apartheid viewpoint. It posited 'mass segregation' as an ideal and envisaged settling the different African tribes in separate areas. In these areas Africans would be able 'to develop and express themselves in the political, economic, cultural, religious and educational spheres'. 'Detribalized urban natives' would be encouraged to settle in these areas. Those who refused would be settled in urban locations where they would enjoy no political or property rights.[20]

Several leading clergymen were members of the Broederbond and they soon won acceptance for the policy within the church. The Dutch Reformed Church (DRC), by far the largest Afrikaans church, was most prominent in its advocacy of apartheid. In 1935 the Federal Council of all the Dutch Reformed Churches formulated a policy with regard to missionary work which would dominate thinking in the church for the next three to four decades and which would, at the same time, provide an ethical basis for apartheid.

The policy, as formulated in 1935, noted the Afrikaners' traditional fear of *gelykstelling* between white and black, and expressed the DRC's strong opposition to any racial mixing. At the same time it declared that it did not begrudge coloured people and Africans a social status 'as honourable as they could reach' and that each nation had the 'right' to be itself, and to develop itself and raise itself. The Federal Council concluded by stating that social differentiation and spiritual separation would be beneficial to both whites and blacks.

In the evolution of DRC thinking, 1935 was a turning point. In later years, the church made little attempt to use the Scriptures as the point of departure when formulating its stand upon social issues. Instead, it took traditional Afrikaner norms as motivation for its decisions. It believed that God created nations and shaped their destinies: the course along which a nation was guided, in other words the 'traditional', was an expression of God's will and was thus in accordance with the Scriptures. As Afrikaner nationalists, the church leaders believed in apartheid and used scattered texts and the history of Israel to provide some moral justification for their actions.

The 1935 policy statement was also important in that it pointed to the growing trend within the church to consider man as part of a collectivity rather than as an individual. Before 1935 the church had emphasized the equal worth of all individuals before God; this now changed into an acceptance of the equal worth of all nations. In later years the DRC condoned individual suffering which resulted from the prohibition of sex across the colour line and forced removals under the Group Areas Act, because these laws were seen to be indispensable to the maintenance of separate nations.[21]

In 1942 the DRC established the Federale Sendingraad (Federal Missions Council) which made a special study of the racial question. It became one of the driving forces behind the founding of the Suid-Afrikaanse Buro vir Rasse-aangeleenthede (SABRA). This united the church leadership and Afrikaner academics in the search for a solution to the racial question in South Africa. In 1943 the Council sent a delegation to government to request not only a strict ban on mixed marriages and extra-marital racial mixing, but also the implementation of segregated residential areas. The following year the same issues were again discussed with the government, but this time the delegation also insisted on the development of separate industries for whites and blacks.

The delegates emphasized that Africans and coloured people had to enjoy 'all possible rights and privileges in their own territories and on their terrain'. The general idea was that they 'should not in the least feel that they were done an injustice but that they could realize themselves and be an asset for the land and their own nation'. Smuts replied that the government was in favour of segregated residential areas and keeping workers of separate races apart, but added that racial sex laws were impractical because a dividing line between whites and coloured people could in many instances never be drawn.[22]

In 1947 a study commission of the DRC Federale Sendingraad further developed DRC thinking on apartheid. It stated:

> DRC policy amounts to the recognition of the existence of races and nations as separate units foreordained by God. This is not the work of human beings. Accordingly the DRC considers it imperative that these creations be recognised for the sake of their natural development through which they could fulfil themselves in their own language, culture and community. Although God created all nations out of one blood He gave each nation a feeling of nationhood and a national soul which had to be recognised by everyone.[23]

The study commission also argued that the best way in which whites and non-whites could coexist was by way of a system of apartheid in which each group developed separately and in its own sphere. Each nation had the right to be itself and not to simply ape other nations. As far as the coloured people were concerned, the commission argued that they would shake off their feelings of inferiority only when they had developed pride in themselves as a nation. This could be facilitated by giving them separate residential areas in which they could get better housing, more say in their own administration, and all the facilities provided to whites. Along this route they would develop the 'greatest sense of responsibility with respect to themselves'. As regards Africans, the commission argued that the DRC support for apartheid did not entail any wish to retard them, but was aimed at 'protecting them against exploitation and ensuring their development in their own territory and their own towns'.[24]

In formulating its views on racial policy, different emphases developed within the DRC. The Free State and the Transvaal churches were more explicitly racist in arguing for a system of racial apartheid which would maintain the racial hierarchy. The Cape church, however, tended to stress the equal worth of each *volk* which, nevertheless, had to develop separately. These different attitudes became evident when the church was challenged by two outstanding DRC theologians, B. J. Marais in the Transvaal and B. B. Keet in the Cape, who questioned whether any Biblical grounds existed for apartheid. At the 1948 Transvaal synod it was argued that the history of Israel of the Old Testament offered a paradigm for racial apartheid. The Cape church, on the other hand, supported a policy of vertical apartheid which, it believed, did not imply inferiority or oppression, but which would enable each group to reach its peak of development. The Cape church then attempted to give Biblical backing to the claim that races and nations (*volke*) were not only permitted by God, but were explicitly decreed by Him.[25]

The Sharpeville massacre of 1960 could have been a turning point but the DRC did not take up the challenge. Delegates from the Transvaal and Cape DRC, who had been deeply shocked by the political events, prepared documents for a consultative conference convened by the World Council of Churches in December 1960. The concluding statement of the conference rejected all unjust discrimination (specifically migrant labour, job reservation and the ban on racially mixed marriages) and recognized all racial groups as part of 'our total population'. It also stated: 'The spiritual unity among all men who are in Christ must find visible expression in acts of common worship and witness, and in fellowship and consultation on matters of common concern'.[26]

The DRC delegates supported the concluding statements, but when the Prime Minister, Dr H. F. Verwoerd, strongly attacked the statement, the Cape and Transvaal synods fell into line, repudiating their own elected representatives. In response, Beyers Naudé, moderator of the Transvaal synod, founded the Christian Institute as an ecumenical organization. Shortly after this he was deprived of his ministerial status by the church.

In the course of the 1960s and early 1970s, the DRC further developed its support for a policy of 'differentiation of peoples'. By 1974, however, the church had come to reject racial discrimination and injustice in principle. In addition, it was no longer prepared to base its support for apartheid on traditional policy or, as some church ministers had done many decades earlier, Old Testament episodes such as the Curse of Ham. In its national synod of 1974, the DRC came up with a somewhat more sophisticated statement which they based on the creation narratives and the proto-history of Genesis, 1 – 11. Two dominant themes emerged in the report. On the one hand, the synod declared that the Scriptures upheld the essential unity of mankind and fundamental equality of all peoples. On the other hand, however, it stated that 'ethnic diversity is in its very origin in accordance with the will of God'. Although the diversity of peoples was relative to their underlying unity, the synod declared that the Bible made provision 'for the regulation, on the basis of separate development, of the coexistence of various peoples in one country'. While it was the duty of the church to ensure that diversity did not lead to estrangement, the synod also warned against 'the modern tendency to erase all distinctions among peoples'.[27]

This formulation left the corner-stones of apartheid by and large unchallenged.

Tenet two:

Individuals could realize their human potential only through identification with and service of the volk; hence, institutions such as the school and the church had to be used to compel an individual to a close identification with his volk.

This belief should be seen against the background of a much larger debate on the relationship between education and culture which had been waged, especially in church and missionary circles, during the first half of the twentieth century. Missionaries active in the field of education had been united in the principle of bringing Christianity and 'civilization' to Africa. However, early in the twentieth century, a few had begun to question the kind of education and civilization they were supposed to introduce. Some had argued that the circumstances and the character of blacks and whites were profoundly different and that the education of each group should be structured accordingly, lest the black man be made a poor imitation of the white man.[28]

This was a question that invited anthropological analysis of the role and value of traditional African structures. It was therefore no coincidence that one of the most influential Afrikaner nationalist thinkers was Werner Eiselen, who taught social anthropology at the University of Stellenbosch. In a pamphlet published in 1929 Eiselen warned about the dislocating effect that migrant labour and urbanization had on traditional African structures. He stated that government policy, and in particular educational policy, was not geared to the gradual development of Bantu culture. Instead, the policy served to promote assimilation between native and European because it took such an extremely negative attitude towards the indigenous culture of the African.[29] In 1934 he again expressed his misgivings about the situation: 'Educated people among the Bantu do not seem to be under any sense of obligation towards their sibs and tribes'. He urged a co-ordinated policy aimed at 'helping the Bantu ... in the building of peoples and cultures'.[30] Twenty years later, when Eiselen became Chairman of the Native Education Commission, he spelt out in great detail his ideas about Bantu education.

By contrast, liberal thinkers in the 1930s and 1940s increasingly advanced the assimilationist argument that African culture had irreversibly broken up under the impact of the Europeans. They believed that Africans could not possibly find fulfilment through identifying only with their own separate cultures and groups. In their view, a modern, secular, scientific society had to be forged in which ethnicity, religion or race played little or no role.

Evangelical circles also debated the relationship between modernization and Christianization. Here, too, the liberal view was that the universal thrust of the Christian faith had swept away differences of race, culture, language and religion. Against this stood those who argued for the preservation of the African culture on religious grounds. One of the influential scholars was the German theologian Diedrich Westermann, who rejected the notion that Christianization necessarily meant Westernization. He castigated those missionaries who treated African peoples as though they had no religion, language or character of their own and who thought that the destruction of all African culture was a precondition for success. In Afrikaner nationalist circles these thoughts were developed into Christian Nationalism which, in essence, argued that Christianity should link up with the national heritage of peoples. In the Christian-National scheme of thought the Christianization of an ethnic group enriched and enobled its national identity.

The debate also extended to the education of coloured people and Africans. During the first three decades of Union the dominance of segregationist thought in education was not seriously challenged. Edgar Brookes formulated the prevailing view: 'If our educational system tends to treat the native as an imitation white man, and our social, economic and political institutions keep on reminding him that he is not a white man, trouble is sure to result'.[31]

While the English-speaking intelligentsia in the 1930s slowly began to move in a liberal direction and toward a common education system, the Afrikaner nationalists turned to apartheid. In 1935 the DRC Federal Council gave one of the first formulations of the Christian-National policy that it hoped to implement with respect to both the coloured people and Africans. Firstly, the DRC wanted education to be in accordance with the principles of the Bible. Secondly, the education system had to prepare coloured people and Africans for the demands of Christian civilization and their own natural environment. In other words, the purpose of education was to enable them to assume their assigned place 'in their land and ethnic group, something they could not do if they were to simply ape whites'. Accordingly, education had to be based on a group's national culture, and had to give proper place to the group's language, history and customs. 'Education must not be denationalized', the policy statement concluded.[32]

The early 1940s saw the first skirmishes between liberals and apartheid ideologues on the education policy towards Africans. Officials in the Smuts administration began to press for a decision on the future course of education policy. In 1944 the Minister for Native Affairs convened a conference to consider alternatives. With few exceptions, the English-speaking delegates all opted for African education to be put under a common department of education. All delegates of the Afrikaans churches strongly protested against such a 'dangerous' policy. They wanted African education to remain under the Department of Native Affairs and be geared towards maintaining the policy of apartheid, and Christian Nationalism in particular. As one delegate put it, 'Inferior it [African education] does not have to be, but it definitely has to be different from that of whites because the native has a lower level of development and has different needs from whites'.[33]

During the 1940s, Afrikaner educationalists spelt out the doctrine of Christian-National education. They argued that the Afrikaners, whose culture and national aspirations had almost been destroyed by British hegemony, had a special obligation to guide the development of Africans in the only positive direction, that of national self-fulfilment. The Christian-National educationalist, B. F. Nel, stated:

> [The] whites in South Africa are responsible for the spiritual awakening of the native and thus also for the direction in which this spiritual energy develops itself in future. This task rests the more heavily on the Afrikaner who came into being and grew up on African soil, and who, through bitter experience, has a deep understanding of the two ways the spirit can take in its development: that is, the 'developmental' direction of the denationalized spirit which is nothing but an imitative spirit, and the truly creative developmental direction which is the direction of the genuine culture creator, the national essence.[34]

By the time the Nationalists came to power in 1948, Christian Nationalism, as a framework for education policy, was in place. As a

Federasie van Afrikaanse Kultuurverenigings (FAK) pamphlet of 1948 expressed it, the Christian principle of this policy meant that education had to be based on the Bible, while the nationalist principle demanded that for all ethnic groups the school should inculcate love of 'their own' and, in particular, a love of their country, language, history and culture.[35]

Soon after coming to power the Nationalists began to restructure education according to the principles of Christian Nationalism. In the case of African education, this included the establishment of a close link between schooling and the reserves or 'homelands'. The Eiselen Commission report of 1951, argued that cultural differences created the need for separate schooling.

> Educational practice must recognize that it has to deal with a Bantu child, i.e. a child trained and conditioned in Bantu culture... The schools must also give due regard to the fact that out of school hours the young Bantu child develops and lives in a Bantu community and when he reaches maturity he will be concerned with sharing and developing the life and culture of that community.[36]

Both Eiselen and Hendrik Verwoerd, who was Minister of Native Affairs from 1950 to 1958, concluded that the urban areas in which 'many cultures are herded together'[37] were not conducive to the orderly development of Bantu cultures. Speaking in 1954, Verwoerd rejected the existing system of African education which prepared children for incorporation into a white society unable to accommodate them:

> There was no place for him [the African] above the level of certain forms of labour. Within his own community, however, all doors are open. For that reason it is of no avail for him to receive a training which has as its aim absorption in the European community while he cannot and will not be absorbed there. Up to now he has been subjected to a school system which drew him away from his own community and practically misled him by showing him the green pastures of the European, but still did not allow him to pasture there.[38]

He went on to say that education should therefore 'have its roots entirely in the Native areas and the Native environment and the Native community ... The Bantu must be guided to serve his own community in all respects'.[39] The Nationalist government announced similar, but slightly more moderate policy goals for coloured and Indian education. Apartheid education became a corner-stone of the South African state and even the changing needs of the economy and society have not deflected government from this policy. In 1980 it asked the Human Sciences Research Council to make recommendations for a feasible education policy that would allow for the realization of the inhabitants' potential and promote economic growth. A representative panel recommended a set of principles which stressed equal opportunities for all and the freedom of choice of the individual, parents and organizations in society. Although the government appeared to accept these principles, including that of the freedom of choice, it added the proviso that each population group was to have its own schools, thereby effectively undercutting the liberal thrust of the report.[40]

Tenet three:

> *Apartheid was the best means of avoiding political friction, while it also ensured the survival of whites and facilitated the highest measure of development for each ethnic group.*

The architects of apartheid fully accepted that wide-ranging state intervention would probably be necessary in order to reverse the process of integration and to induce ethnic groups to live together and exercise their political rights through established apartheid channels. National Party politicians justified massive state incursion into the lives of individuals by stating that numerous other states also used compulsion to promote amalgamation in order to ensure justice for all. The NP did not believe in amalgamation. It maintained, however, that its own ultimate aim of apartheid was nevertheless as 'exalted' and much more progressive than segregation, which Malan had dismissed as mere 'fencing off' in 1944.[41] In 1948, Diederichs argued that although apartheid differed radically from liberalism, it was not necessarily inferior in terms of its premises and goals:

> [W]hat is at issue is two outlooks on life, fundamentally so divergent that a compromise is entirely unthinkable except on superficial matters .... On the one hand we have nationalism, which believes in the existence, in the necessary existence, of distinct peoples, distinct languages, nations and cultures, and which regards the fact of existence of these peoples and these cultures as the basis of its conduct. On the other hand we have liberalism, and the basis of its political struggle is the individual with his so-called rights and liberties.[42]

Apartheid was also considered to be ethically justifiable because it prevented friction and conflict. A Stellenbosch academic, N. J. J. (Nic) Olivier, who emerged as one of the most influential apartheid thinkers, wrote:

> That such a policy, which aims at forestalling the race conflict that is inherent in the present situation by removing the root cause of the problem – the intermixture of races – can be justified on Christian, ethical and scientific grounds is clear.[43]

Behind Olivier's view lay the Nationalist conviction that their policy offered a better formula for dealing with the explosive forces within South African society than the one proposed by the liberals. Blacks and whites living together in integrated residential areas and competing in a common system for power, jobs, access to amenities and services would, in the Nationalist mind, inevitably lead to escalating conflict. The greater the number of competitive points of contact, the greater the friction. Hence the apartheid formula: reduce the points of contact by increasing separation at all levels of society, be they political, economic or social. Olivier defined apartheid's main policy aim with respect to the relationship between whites and Africans as the 'gradual and systematic disentanglement of the two groups, making it possible for each group to exercise political rights and enjoy economic opportunities within its own territory'.[44]

To apartheid advocates, state-enforced separation was necessary to ensure the survival of whites. Only through apartheid could whites maintain themselves in a society in which they were outnumbered and in which there existed fundamental differences between white and black in civilization, culture, and a general way of life. They believed that failure to impose apartheid would lead to the political suicide of whites. Consequently, Africans should exercise and have rights only in their own areas and communities. Economically, white workers should be protected in the labour market to enable them to maintain their living standards. Socially the policy should aim at maintaining the racial differences.[45]

Among apartheid ideologues different approaches developed with respect to the extent of separation. There were, on the one hand, academics, such as Olivier, and professional people drawn together in the Federasie van Afrikaanse Kultuurverenigings (FAK) the front of the Afrikaner Broederbond, and the Stellenbosch-based Suid-Afri-kaanse Buro vir Rasse-aangeleenthede (SABRA), who increasingly argued the case for radical separation. This meant the creation of 'national homelands' for Africans in which the great majority of Africans were supposed to find political, economic and cultural expression. This entailed the systematic removal of Africans to the homelands until only a few were left in the white homeland. In principle, the very presence of Africans in urban areas, regardless of whether or not they were employed by whites, was undesirable. These members of the Afrikaner intelligentsia believed that South Africa, as a white country, was politically doomed if the economy became increasingly dependent on black labour. Moreover, they argued that it would be immoral to continue to deny political rights to people who were becoming increasingly integrated into the economy and society; hence they demanded the dynamic economic development of the homelands in order to justify sending increasing numbers of Africans back to these areas. The Tomlinson Commission report of 1956 gave a detailed exposition of the financial implications of a policy of radical separation, but this was studiously ignored by Verwoerd after his assumption of power in 1958.

On the other hand, there was the NP political leadership, together with organized Afrikaner business and agricultural interests, who opposed radical partition. They wanted a high economic growth rate and an increased, but differential, incorporation of Africans into the workforce. They believed that Africans could be controlled if they were denied political and industrial rights. They justified this approach with examples of similar policies in countries such as Germany and Switzerland, which employed large numbers of migrant labourers without granting them political rights.[46]

But the differences between the idealistic Afrikaner intelligentsia and the more pragmatic party leadership should not cloud the fact that, within the NP fold, a general consensus existed about the Afrikaner's historical right to the land and political sovereignty. It was expressed unambiguously by Prime Minister Vorster: 'We have our land and we alone shall have the say over that land. We have our Parliament and in that Parliament we alone shall be represented'.[47] This principle was enshrined as more homelands were given 'independence' and Africans were deprived of their South African citizenship.

Tenet four:

Apartheid granted others that which one group (the Afrikaner) demanded for itself.

This argument was made by both religious leaders and by secular intellectuals. From the 1940s, the DRC had insisted that its support for apartheid in no way detracted from its commitment to justice, freedom and prosperity for all. In fact it believed that these ideals could only be achieved through apartheid and that blacks would only become free once they were self-reliant; in the DRC's view apartheid gave expression to the Christian ethic of 'grant unto others' that which one (the Afrikaner) demanded for oneself.

After an exhaustive study of DRC documents on apartheid, Kinghorn remarked that the words 'we grant' appeared constantly. For instance, if the DRC demanded separate residential areas for whites, it was prepared to 'grant' to the various black groups the same. If Afrikaners (and the larger white community) wanted freedom, it 'granted' others the same. In this respect the DRC was keen to bolster its argument by distinguishing between horizontal and vertical separation. The church portrayed segregation as being a form of repression because it was based on a horizontal separation, which meant that people were divided into two classes, that is, master and servant, or upper and lower class. By contrast, apartheid was presented as a form of vertical separation, which put the various groups not above (or below), but alongside one another. According to the church, this was the truly Christian way of coexistence. This form of coexistence did not make all individuals equal to one another; however, the various groups would be 'equal in their freedom'.[48]

Secular intellectuals were more concerned with finding a true political principle, which was true for everyone and which was not in conflict with the principles on which the states of Western Europe rested. These intellectuals grappled with the challenge posed by R. F. A. Hoernlé, a brilliant liberal thinker of the time. He distinguished between segregation, which he saw as nothing more than an instrument of domination, and 'total separation' which envisaged 'an organisation of the warring sections into genuinely separate self-contained, self-governing societies, each in principle homogeneous within itself, which can then cooperate on a footing of mutual recognition of one another's independence'.[49]

In response to Hoernlé, the oustanding Afrikaans writer, N. P. van Wyk Louw, rejected the application of the classic liberal model of decolonization which insisted on various peoples sharing freedom in one country and within one nation. This, to Louw, constituted an injustice in the 'multi-national' South African state. As a solution he proposed a liberal nationalism, which meant that neither liberalism nor nationalism should be rejected. The cardinal idea which he advanced was that 'the recognition of the nationalist principle for the Afrikaner logically has to imply the recognition of all other national groups in South Africa'. Hence, the Afrikaners could not speak of themselves as 'the' *volk* of South Africa but merely as one of the *volke* of South Africa.[50]

Further thinking in this mould appeared in a column of the Cape Town newspaper, *Die Burger*, in which two gifted journalists and essayists, Piet Cillié and Schalk Pienaar, developed the idealistic strand of apartheid. Cillié put forward what he called the 'principle of separate freedoms', something which he claimed developed out of the apartheid idea 'once it was recognized that in principle no artificial limits could be set for Bantu self-government in the homelands'.[51]

The politicians were not slow to take up this theme and use it to justify party policy. In 1950 Verwoerd, in a speech to the Natives Representative Council, stated:

> The present government adopts the attitude that it concedes and wishes to give to others precisely what it demands for itself. It believes in the supremacy [*baasskap*] of the European in its own sphere but, then, it believes equally in the supremacy of the Bantu in his own sphere.[52]

It was the idea of separate nations and of separate freedoms which increasingly shifted the NP away from the crude racist thought which had prevailed in the early years of the Nationalist government. Since the mid-1950s, Nationalist politicians have presented NP policy as being moral because it is based on the same mobilization principle which had achieved such great things for the Afrikaners.

When De Wet Nel addressed a SABRA congress on the goals of apartheid, he used language which, without changing a word, could have been employed during the 1930s at a rally to propagate Afrikaner ethnic mobilization.

> The personal national ideals of each individual and each ethnic group could best be realized within his own community in his own ethnic group. In other words ethnic development needs to be ethno-centric. Each *volk* in the world is entitled to the best efforts [*kragte*] of its own sons and daughters. A policy which is designed to rob a *volk* of this is immoral. Such a policy is in fact the wasteful exploitation of people and has always been one of the most important sources of racial hatred.[53]

The NP also increasingly insisted that apartheid did not imply that blacks were inferior, merely that they were different. In 1960, De Wet Nel challenged his audience to name a single instance in which a white leader had claimed that blacks were doomed to permanent racial inferiority.

By the mid-1960s Nationalists were expressing the ethic of 'grant unto others' in grandiloquent terms. The Afrikaners' mission as a 'mature *volk*' was seen to be the rescue of other demoralized and denationalized *volke*. This mission would be completed only when the other *volke* and communities had accepted self-determination, which included complete independence. This ethos of apartheid guardianship was succintly expressed in the preamble to the Prevention of Improper Interference Bill of 1966, which prohibited all political organization across the lines of the defined ethnic groups:

> Whereas the traditional way of life of the Republic of South Africa requires that every population group shall develop independently within its own group, but with mutual co-operation and assistance; and whereas every population group has an inalienable right to live and to strive according to its own traditional way of life, as being the only foundation for ensuring lasting peace and good order, and whereas the whites as the guardians of the other population groups accept their mission to lead the non-white population groups to self-realization and to safeguard them against political exploitation by others as the sole guarantee for the continued existence of both their own and the other population groups...[54]

The apartheid ideology thus recast liberal values in nationalist terms and applied them to South Africa. The country was projected not as a nation-state, but as a region comprised of several national states and national communities. Liberty was defined as national self-determination, and equality as equal full nationhood. It was envisaged that people would attain political rights in their respective homelands or, in the case of the coloured people and Indians, in their respective communities.

Conclusion

Ideology has two dimensions, fundamental and operative. Afrikaner nationalism emphasized the fundamental aspect — the goal of national independence — but was weak on the operative dimension. Apartheid, by contrast, has both a fundamental dimension (self-determination of peoples and separate freedoms) and a well-developed operative dimension. The latter includes the principles which guided concrete political action and which found expression in separate residential areas, schools, and public amenities.

Although apartheid was conceived of as being the mere instrument of Afrikaner nationalism, it tended to assume a hegemonic position during the 1960s, when it was taken to its extreme in prescribing public behaviour and private intercourse and in regulating the position of blacks in society. Both the fundamental and operative dimensions were necessary in order for apartheid to serve as the ideology of Afrikaner unity. Without the fundamental dimension (sometimes called 'grand apartheid') which held up the ideal of 'separate freedoms', the ideology would not have attracted the support of often well-meaning Afrikaner intellectuals, who would not have countenanced mere racism. But the operative dimension was equally indispensable in persuading the ordinary voter that the government was constantly restructuring society to ensure white security. It was primarily Verwoerd who developed apartheid thinking into a coherent ideological system.

From the late 1960s, Afrikaner unity began to crumble. This was a result of both a crisis in the apartheid ideology and the class divisions within Afrikaner society. Apartheid ideology started to disintegrate when the homelands as vehicles for 'separate freedoms' began to lose credibility. At the same time, lower-class Afrikaners became convinced that their interests were being threatened by growing integration which was resulting in blacks climbing ever higher on the job ladder and encroaching on their privileged positions.

Various intellectuals who had originally helped to formulate the apartheid ideology started to break away from the NP for very different reasons — thus highlighting the distinct strands out of which apartheid was woven. For instance, N. J. J. Olivier, who broke away to join a liberal party to the left of the NP, was representative of the group of Afrikaner nationalists who propagated a 'liberal nationalism' similar to that proposed by Van Wyk Louw. Olivier found it impossible to condone the fact that apartheid had become a cloak for massive discrimination and injustice. In contrast, others such as Albert Hertzog and Andries Treurnicht, who formed conservative parties to the right of the NP, were staunch nationalists who did not approve of the watering down of apartheid. Both pledged that they would restore the Calvinist principles and work for the viability of the *volk*. Once Treurnicht had drawn sufficient support to challenge the government, however, he began to downplay the emphasis on Calvinism and Afrikaner exclusivity. He now argues that full-scale apartheid is strategically necessary to ensure white survival.

President P.W. Botha was an Afrikaner nationalist who was prepared to shed some of the non-functional aspects of apartheid. Nevertheless, he was still driven by the belief that ethnic groups, as statutorily defined, were the chief 'natural' characteristic of South African society. In his view, it was almost a natural law that an individual's classification within an ethnic group should determine where he goes to school, in which residential area he may live, and through which political structures he may exercise political rights.[55] Where he differed from Treurnicht was in the strategic concessions he was prepared to make to win political allies across racial lines and to regenerate the economy. But the apartheid ideology which took shape from the mid-1930s remained a dominant element in his convictions.

Notes

1. A. James Gregor, *Radical Ideologies*. New York: Random House, 1968.
2. N. J. Rhoodie and H. J. Venter, *Apartheid*. Cape Town: HAUM, 1959 and also N. J. Rhoodie, *Apartheid and Racial Partnership in Southern Africa*. Pretoria : Academica, 1969.
3. Gregor, *Radical Ideologies*, p. 240.
4. Gerrit Viljoen cited by John de St Jorre in *A House Divided*. New York: Carnegie, 1977, p. 13. See also Gerrit Viljoen, *Ideaal en Werklikheid*. Cape Town: Tafelberg, 1978.
5. Carl J. Friedrich, *Man and his Government*. New York: McGraw Hill, 1963, pp. 89 – 90.
6. This distinction is made by George Rude in *Ideology and Popular Protest*. London: Lawrence and Wishart, 1980.
7. Schalk Pienaar and Anthony Sampson, *South Africa : Two Views of Separate Development*. New York: Oxford University Press, 1960, p. 5.
8. D. F. Malan, *Glo in u Volk : Dr D. F. Malan as Redenaar* (compiler Schalk Pienaar). Cape Town: Tafelberg, 1964, pp. 131 – 7.
9. M. C. de Wet Nel, 'Waarom die beleid van apartheid?' *Journal of Racial Affairs*, vol. 11, no. 4, 1960, p. 180, 184.
10. I. Hexham, *The Irony of Apartheid : The Struggle for National Independence of Afrikaner Calvinism against British Imperialism*. New York: Edwin Mellen Press, 1981.
11. T. Dunbar Moodie, *The Rise of Afrikanerdom*. Berkeley: University of California Press, 1975.
12. G. J. Schutte, 'The Netherlands : Cradle of Apartheid?', *Ethnic and Racial Studies*, vol. 10, no. 4, 1987.
13. Moodie, *The Rise of Afrikanerdom*, p. 70.
14. A. P. Treurnicht, *Credo van 'n Afrikaner*. Cape Town: Tafelberg, 1975, pp. 1 – 10.
15. Schutte, 'The Netherlands : Cradle of Apartheid?'
16. Nico Diederichs, *Nasionalisme as Lewensbeskouing*. Bloemfontein: Nasionale Pers, 1936, pp. 17 – 18.
17. Nel, 'Waarom die beleid van apartheid?', p. 173.
18. Geoff Cronjé authored three books: *'n Tuiste vir die Nageslag : Die Blywende Oplossing vir die Suid-Afrikaanse Rassevraagstuk*. Stellenbosch: Pro Ecclesia, 1945; *Regverdige Rasse-apartheid*. Stellenbosch: CSV, 1947 and *Voogdyskap en Apartheid*. Pretoria: Van Schaik, 1948.
19. Cronjé, *Regverdige Rasse-apartheid*, pp. 136 – 7.
20. A. N. Pelzer, *Die Afrikaner-Broederbond : Eerste 50 Jaar*. Cape Town: Tafelberg, 1979, pp. 158 – 9.
21. The two paragraphs preceding the reference are based on an excellent new study by J. Kinghorn (ed.), *Die NG Kerk en Apartheid*. Johannesburg:

MacMillan, 1986, pp. 86 – 9. For the original document, see Nederduits Gereformeerde Sendingkerk : Federale Sendingraad-notules, 1935, pp. 94 – 9.

22. NGK : Federale Sendingraad-notules, 1944, pp. 3 – 4.
23. NGK : Federale Sendingraad-notules, 1947, pp. 137 – 41.
24. NGK : Federale Sendingraad-notules, 1947, p. 141.
25. Kinghorn, *NGK en Apartheid*, pp. 109 – 15.
26. Leslie A. Hewson (ed.), *Cottesloe Consultation : The Report of the Consultation.* Johannesburg: South African Institute of Race Relations, 1961.
27. John de Gruchy, *The Church Struggle in South Africa.* Cape Town: David Philip, 1979, pp. 71 – 2.
28. My discussion of this debate is largely based on John David Shingler, 'Education and political order in South Africa', Ph.D. dissertation, Yale University, 1973, pp. 155 – 79
29. W. Eiselen, *Die Naturelle-vraagstuk.* Cape Town: Nasionale Pers, 1929, p. 8.
30. W. Eiselen, 'Christianity and the religious life of the Bantu', I. Schapera (ed), *Western Civilization and the Natives of South Africa.* London: Routledge, 1934, pp. 81 – 2.
31. Edgar Brookes, *The History of Native Policy in South Africa from 1830 to the present day.* Cape Town: Nasionale Pers, 1924, p. 464.
32. NGK: Federale Sendingraad-notules, 1935, p. 98.
33. P. J. S. de Klerk, 'Die administrasie en finansiering van naturelle-onderwys' in *Koers*, vol. 11, no. 5, 1944, p. 173.
34. Cited by Shingler, 'Education and political order', p. 262.
35. Federasie van Afrikaanse Kultuurverenigings, *Christelike Nasionale Onderwysbeleid.* Johannesburg: FAK, 1948.
36. Report of Commission on Native Education. Eiselen Report, UG 53/1951, paragraphs 773 – 4.
37. Report of Commission on Native Education, para. 770.
38. H. F. Verwoerd, *Verwoerd Speaks : Speeches 1948 – 1966* (compiler A. N. Pelzer). Johannesburg: APB, 1966, pp. 83 – 4.
39. Verwoerd, *Verwoerd Speaks*, p. 83.
40. Human Sciences Research Council, *Provision of Education in the RSA* (De Lange panel). Pretoria: HSRC, 1981; Republic of South Africa, *Interim Memorandum on the Report of the Human Science Research Council on the Inquiry into the Provision of Education in the RSA*, 1981, p. 3.
41. Rhoodie and Venter, *Apartheid*, p. 175.
42. *HAD*, 1948, cols 500 – 1.
43. N. J. J. Olivier, 'Apartheid : A slogan or a solution?', *Journal of Racial Affairs*, vol. 5, no. 2, 1954, pp. 29 – 30.
44. Olivier, 'Apartheid', p. 28.
45. Olivier, 'Apartheid', p. 27.
46. Deborah Posel, 'The meaning of apartheid', unpublished paper, University of London ICS seminar.
47. B. J. Vorster, *Geredigeerde Toesprake van die Sewende Eerste Minister van Suid-Africa, vol. I : 1953 – 1974* (compiler O. Geyser). Bloemfontein: Instituut vir Eietydse Geskiedenis, 1976, pp. 102 – 3.
48. Kinghorn, *NGK en Apartheid*, p. 97.
49. R. F. A. Hoernlé, *South African Native Policy and the Liberal Spirit.* Johannesburg: Witwatersrand University Press, 1945, p. 186.
50. N. P. van Wyk Louw, *Liberale Nasionalisme*, 1958, reprinted in *Versamelde Prosa.* Cape Town: Human and Rousseau, 1986, pp. 502 – 6.
51. Dawie (pseudonym for columnist), *Bloemlesing uit die geskrifte van Die Burger se politieke kommentator* (compiler Louis Louw). Cape Town:

Tafelberg, 1965, pp. 217, 285 – 90.
52. Verwoerd, *Verwoerd Speaks*, p. 25.
53. Nel, 'Waarom die beleid van apartheid?', pp. 171 – 2.
54. Cited by R. de Villiers, 'Afrikaner Nationalism', Monica Wilson and Leonard Thompson (eds), *Oxford History of South Africa, vol. II*. Oxford: Oxford University Press, 1971, p. 414.
55. Koos van Wyk and D. J. Geldenhuys, *Die Groepsgebod in P. W. Botha se Politieke Oortuigings*. Johannesburg: Randse Afrikaanse Universiteit, 1987, pp. 46 – 7.

CHAPTER 3

The functioning of apartheid

There is a widespread assumption that by 1948 a grand apartheid blueprint existed which the National Party (NP) then systematically and progressively implemented.[1] However, a completely different position has been articulated by Piet Cillié, editor of *Die Burger* from 1954 to 1978 and, next to Verwoerd, perhaps the most outstanding intellect in the Afrikaner nationalist movement. In his assessment of apartheid, written in 1985, Cillié remarks:

> A system? An ideology? A coherent blueprint? No, rather a pragmatic and tortuous process aimed at consolidating the leadership of a nationalist movement in order to safeguard the self-determination of the Afrikaner.[2]

Both perceptions have to be taken into account in order to understand the implementation and functioning of apartheid as a system. As outlined in the previous chapter, apartheid was an idea that could readily be applied to virtually every concrete situation. And because both the ideology and praxis of apartheid appeared to safeguard Afrikaner rule, there was no shortage of politicians willing to apply it. On the other hand, in the 1950s, the NP lacked the means and conviction with which to impose neat political and social solutions upon a society that was steadily becoming more integrated. Inevitably this gave rise to major political contests and struggles — between the government and blacks, between the government and organized business, and between forces within the Afrikaner nationalist alliance. At all times the overriding principle was not apartheid as a goal in itself but whether a particular policy decision was in the interest of Afrikaner 'self-determination'.

One can identify two distinct phases of apartheid. During the first decade after 1948 the Nationalist government purposefully introduced apartheid's statutory groups and compulsory residential segregation. But in other areas, such as regulating labour, curbing African urbanization and stamping out black opposition, its efforts were hesitant and contradictory.

It was a different story in the 1960s. H. F. Verwoerd, who became Prime Minister in 1958, was a determined and uncompromising ideologue. He had the ability and, above all, the will to impose apartheid as a systematic policy over the efforts of his opponents within the party who spoke of pragmatic adaptations and exemptions. Moreover, Verwoerd acted from a position of strength. During the early 1960s the state had managed to suppress the upsurge of black resistance with unexpected ease and the fear of effective international intervention had proved groundless. Most importantly, the 1960s saw a great economic boom which put large resources in government hands. Apartheid, which was at this stage presented by the NP as separate development, intensified. All vestiges of black representation in the 'white' political system were removed. The government rigidly applied influx control and other curbs on African urbanization. It moved great numbers of people out of white group areas and from 'black spots'.

Looking back on twenty years of Nationalist rule, Prime Minister John Vorster declared in 1968 that apartheid had brought peace to the land. It had done so by 'recognising the right of existence of distinct nations and colour groups', by eliminating points of friction between them and 'by providing each with opportunities to develop according to their ability and with the maintenance of their identity'.[3] This was apartheid at its apogee. The ideological smugness of the NP leadership was almost boundless.

The apartheid system comprised the following main elements: *labour regulation* (influx control and other forms of regulating labour); *communal apartheid* (statutory race classification, group areas, separate education, separate amenities and other residual forms of apartheid); and *political control and privilege* (an all-white Parliament, separate homelands for Africans and discriminatory spending on social welfare).

Labour regulation

After the rapid economic integration of South Africa in the decade before 1948, the NP was immediately confronted with the problem of devising a system which would use and guarantee sufficient African labour without fatally undermining Afrikaner and wider white supremacy. The issue of African labour controls loomed so large that many analysts, particularly Marxists, considered the regulation and reproduction of cheap black labour to be the essence of apartheid. Further developed, this argument holds that the main apartheid innovation was to extend migrant labour, already institutionalized on the mines, to the manufacturing and commercial sectors. A corollary of this view is that the apartheid state managed so successfully to furnish an adequate supply of docile labour that little conflict existed between the capitalist order and apartheid.

These perspectives greatly stimulated the academic debate on apartheid, particularly during the 1970s, but today few scholars support these views without major qualifications. This section briefly addresses the impact of apartheid on the labour order, firstly by examining the restructuring of the influx control system as a whole, before discussing the supply and regulation of labour in the mining, farming and manufacturing sectors.

Influx control

Apartheid met a formidable challenge in the secondary and tertiary sectors and, in particular, in the manufacturing industry with its growing need for a stable and skilled work-force. Manufacturing's contribution to the total national economic output increased considerably in the first two decades of NP rule. In 1950 its contribution to the national economy stood at 18,3 per cent and was less than two-thirds of the combined figure for mining and agriculture; by 1970 the manufacturing sector's contribution had risen to 24 per cent — a third higher than the total for mining and agriculture. At the same time, there was a massive increase in the number of Africans employed in the manufacturing and commercial sectors.

Although the incoming NP government's chief nightmare was that rapid African urbanization would fatally undermine white supremacy, it was also concerned that the flow of labour to the cities would bypass the farms. During World War II and in the years immediately following it, the farmers experienced an acute labour shortage. The swing of the farmers' vote was crucial to the NP in its electoral success of 1948.

The Nationalist government immediately tightened up the system of influx control.[4] As outlined in the first chapter, the ideological foundation of influx control dated back to the Stallard Commission of 1922, which decreed that an African should only be in the urban areas (deemed essentially the white man's creation) to minister to the needs of the white man and 'should depart therefrom when he ceased to minister'.[5] The Native Urban Areas Act of 1923 constituted the legal basis of influx control. In 1934 adult African males throughout the country were compelled to carry passes (previously this was restricted to the northern provinces). Influx control was elaborated through legislation passed in 1937 and 1945. It was decreed that Africans from the reserves were only allowed in the cities for a limited period in order to seek work; they were not allowed to bring their families with them and redundant workers were threatened with expulsion to the reserves. Employers had to register all service contracts made with male Africans. In 1952 the NP passed the Bantu Laws Amendment Act which further toughened up the laws. South Africa now had one of the most rigid influx control systems in the world.

What were the main differences between influx control before 1948 and the system introduced by the NP? First the apartheid system was tied to much more ambitious political and ideological objectives. Nationalist spokesmen expressed the view that influx control together

TABLE 3.1 AFRICAN EMPLOYMENT IN MANUFACTURING AND COMMERCE

Year	Manufacturing	Commerce	Total (Africans)	All population groups in manufacturing and commerce
1950	267 070	69 281	336 351	843 326
1960	357 700	138 240	495 940	1 055 457
1970	617 200	186 300	803 500	1 615 700

Source: Stanley Greenberg, *Race and State in Capitalist Development*. New Haven : Yale University Press, 1980, pp. 425 – 6.

with a greater number of job opportunities in the homelands would, by 1978, produce a situation where the flow of Africans from the homelands to the cities would start to reverse itself.

Second, influx control under NP rule was much more comprehensive and stringent than before. The 1952 Act, together with a 1957 amendment, introduced a rigid division of labour between the low-paying primary sector and the other sectors of the economy. Restrictions on farm labourers and migrants ensured that the mines and farms faced no real competition from the better-paying manufacturing sector. Whereas legislation before 1952 had basically regulated African entry into cities, the new system lay down the conditions under which Africans could remain in the cities. Prior to 1952 some categories of educated Africans, such as church ministers and teachers, had been exempted from the pass laws. However the new system obliged all Africans (including women from 1957) to carry a single reference book. This reference book replaced the multiple documents, such as the poll tax receipt and pass book, which Africans had previously been forced to carry.

Thirdly the new system laid down a much more restricted period in the urban areas for the various work seekers. The 1937 law had allowed a fortnight; this was now cut down to seventy-two hours. The 1952 law also introduced the concept of certain categories of Africans who were exempted from the blanket prohibition on all Africans to remain in a prescribed area for more than seventy-two hours; however, none were exempt from carrying a pass book. In effect, every urban area in the country, excluding those in the homelands, fell within one or more of the prescribed areas.

From 1952 to 1986, when pass books were abolished, Africans generally called the exemptions from the blanket prohibition Section 10 'rights'. At best, these were only qualified rights contingent upon an individual's ability to find lawful accommodation in a particular urban area. Section 10 permitted the following categories of Africans to be in the cities: (a) persons who had since birth continuously resided in the prescribed area; (b) persons who had worked continuously in the area for one employer for at least ten years, or who had lawfully resided in the area with more than one employer for at least fifteen years; and (c) the wife and children of those people who qualified under the previous two categories. A fourth category regulated the position of migrants and commuters from the homelands, and conferred no continuing right on an individual to remain in the area. If the labour bureau's permission lapsed or was cancelled the person concerned had to leave the 'prescribed area'.

However, the pass laws were only part of a larger system of influx control. Other forms of control included:

- *Allocation of labour*: As the system evolved it became increasingly difficult for Africans from the homelands to find employment without which they could not legally remain in the towns. Africans holding Section 10 'rights' in a particular town did not have to register as work seekers, and were given first claim to jobs in that town; second in line were Africans resident in a specific region, and only third, migrants from the homelands. From the 1970s on, at least half of the migrant labour-force was without work for long periods.

- *The regulation of employment*: It was an offence to employ an African in a prescribed area without written permission from a labour bureau. This law was frequently transgressed by Africans who, out of desperation, bypassed the labour bureaux, and were prepared to accept the low wages on offer to illegals.[6]
- *Housing controls*: There was a strong tendency among officials to give permits for township housing only to those men with Section 10 exemptions who had dependents living with them. From the late 1950s the government severely cut back on the construction of houses in the townships. It attacked the thirty-year leasehold which defined the terms of African home-ownership and pressed local authorities to purchase leasehold property on the expiry of leases. In 1968 it announced that new accommodation for Africans families would not be provided in townships in 'white' areas. For much of the 1970s it was policy to concentrate on building houses in dormitory towns in the homelands from which commuters could travel to work. The government also imposed a freeze on all housing in townships located close to a homeland.

One should bear in mind that during the first ten to fifteen years of NP rule, influx control was not a fully designed policy ready for implementation. The system evolved mostly through *ad hoc* adaptations which reflected the outcomes of struggles on the ground.

During the 1950s a major contradiction existed within NP policy. On the one hand the government had committed itself to a 'whitening' of the cities and the Stallard principle which stated that employment was the only justification for the residence of Africans in the cities. On the other hand, economic growth forced the government to meet the labour needs of employers which resulted in a steady increase in the number of urban Africans during the 1950s.

In order to solve this contradiction the government tried to make a clear distinction between various categories of African workers. Migrant workers were considered to be temporary and rightless, and the government wanted to steadily reduce their numbers by restricting employment opportunities for them in the urban areas. The second category comprised the urbanized and stabilized urban Africans whose exempted status under Section 10 meant that they were legally in the cities, regardless of whether or not they were employed. In the government's view, urbanized Africans had to be trained so that they could move up to the semi-skilled positions after whites had been promoted to skilled or white-collar jobs.

The government wanted employers to give preference to Africans exempted under Section 10; migrant labour should be employed only once the existing urban reservoir had been exhausted. Employers, however, preferred the docile and cheap migrant labourers to urbanized Africans, who were considered to be choosey, less disciplined and expensive. This tendency was reinforced by the fact that industrial enterprises during the 1950s were relatively small-scale and labour intensive and showed little inclination to mechanize.

The employers generally got their own way. By the end of the 1950s the government's original intentions had largely been thwarted. The size of the urban African population increased from 2,4 million in 1951 to 3,5 million in 1960, a growth of 45 per cent, and the number of Africans employed in the manufacturing and commercial sectors increased by almost 50 per cent. The 1950s also saw an increase in migrant labour. On census day in 1960 there were 65 per cent fewer migrant workers in the rural areas than there had been on the day the 1946 census was taken.

During the 1960s the government made a determined attempt to reverse these trends. To start with, it removed the administration of large townships from the white city councils and placed it under the Department of Bantu Administration and Development (BAD). BAD became ubiquitous in its the control over virtually every aspect of the daily lives of Africans — their movement (passes), labour, housing, education and social welfare. In his book *Against the World*, the best portrayal of the 1960s, Douglas Brown wrote:

> The Department of Bantu Administration and Development is what the Colonial Office was to Britain a century ago — if you can imagine the people over whom it ruled milling around in the streets of London. There is something uncanny about this mighty *imperium in imperio*, which must be one of the biggest and most all-embracing administrative machines on the face of the earth. To the whites it is invisible, but to the blacks it is ever-present ... No dictator in history has exercised more direct power over human beings than the South African Minister of Bantu Administration.[7]

The government also launched a concerted attack upon the rights and claims of urbanized Africans. It began to make life for urbanized Africans as uncomfortable as possible in an attempt to make them identify with the homelands. In the 1950s the government had accepted that a considerable part of the African population had become urbanized and that their numbers would grow; now it attempted to freeze and even curb the growth of those Africans who qualified to be in the cities. The most drastic move was a change in the Bantu Labour Regulations in 1968 which compelled migrants to return to their homeland at least once a year. This meant that migrants could never qualify for exempted status through being in continuous employment for ten years. In 1970 the government undermined the position of children of exempted people (see p. 99). Even professionals and businessmen whose labour was not strictly required by white employers experienced pressure to move to the homelands (see p. 92).

In short, the 1960s saw a return by government to the Stallard principle of allowing only those Africans catering to the needs of whites in the urban areas. Government spokesmen made no secret of their intention to treat the entire African labour-force as migrants and to remove all superfluous Africans from the cities, regardless of whether or not they qualified for exempted status. In the crudest statement of this policy, a BAD Deputy Minister, Froneman, declared that of the six million Africans in the white urban areas, only two million were economically active; the rest were 'surplus appendages' who should be deported to a homeland.[8]

The government also tried to achieve its objectives by manipulating industrial expansion. Through both incentives and coercion the government hoped to shift approximately one quarter of all new manufacturing investment to designated growth points. Its main instrument was the Physical Planning and Utilization of Resources Act of 1967. This stipulated that ministerial approval was required for the expansion of the African work-force in industry. Permission would only be given if industries were locality bound or had a white to African labour ratio of less than 1 to 2,5. For businessmen this was a serious matter: it was estimated that in the Pretoria-Witwatersrand-Vereeniging area, nearly half of the industries exceeded the 1 to 2,5 ratio.[9]

Between 1968 and 1978 formal applications for the employment of an additional 107 674 Africans in the 'white' areas were refused. The government also used other powers to reject applications for new industrial zones, and to impose new restrictions on the use of existing industrial sites. This prevented another 220 000 African workers from being employed.[10]

Finally the government tightened up controls at the rural end. This it did by greatly expanding the rural labour bureaux to process migrant labour. A migrant could only enter into a labour contract through a labour bureau, and had no opportunity to negotiate the terms of the contract. He could be compelled to take a job in a low-paying sector or area if he did not want to remain unemployed. Once a farm worker, he could find it impossible to be reclassified. As a contract was valid for only a year, a migrant worker had to make sure he satisfied his employer sufficiently for the latter to renew his contract after the mandatory termination of employment. These revamped labour bureaux checked African urbanization until the early 1980s. By this stage the situation in the homelands had become so desperate that work seekers bypassed these bureaux and squatted in townships or on the perimeters of cities. With the abolition of the pass laws in 1986 the labour bureaux largely lost their controlling function.[11]

Influx control was most rigorously implemented in the western part of the Cape Province which many Nationalists saw as the final refuge for whites — a permanent white 'homeland'. In 1957 the government declared the areas west of Aliwal North and the line of the Fish and Kat rivers (that is, most of the western and southern Cape) a Coloured Labour Preference Area. Africans could only be employed in these areas after available supplies of coloured labour had been exhausted.

This Coloured Labour Preference Policy (CLPP), as it became known, also built on precedents from the past. From the beginning of the century coloured people had been in a position of relative privilege compared to Africans in the Western Cape. The coloured people were not subject to influx control (although the shortage of housing in Cape Town acted as a curb) and were traditionally appointed over Africans in semi-skilled positions. As discrimination against the coloured people intensified in the 1930s and early 1940s both white liberals and the Coloured Advisory Council argued that, as a form of quid pro quo, coloured workers should be protected from being undercut by African migrant workers. In 1945 the Cape Town municipality began restricting the issuing of work passes to Africans.[12]

In typical fashion the NP turned hesitant compromises with the racial order into a firm principle and tried to base it on some ethical foundation. The NP argued that Africans had no historical claim to the Western Cape and could therefore be removed without qualms. They also held up intensified influx control as a prerequisite for increased socio-economic opportunities for coloured people. Verwoerd thought that the CLPP offered a semi-privileged status that would reinforce a separate coloured identity and perhaps even foster a coloured 'nationalism'. The latter failed to materialize, but the policy did help to perpetuate the divisions within the black working class of the Western Cape.

In terms of the CLPP, the government strictly controlled the movement of Africans to the Western Cape and expelled redundant Africans. By 1962 over 30 000 Africans had been endorsed out of the Western Cape. In 1964, legislation was passed which permitted the employment of African labour in the Coloured Labour Preference Area only when no coloured workers were available at wages deemed by officials to be attractive enough to draw such labourers. Government spokesmen committed themselves to reducing the number of Africans in the Western Cape by 5 per cent every year. However, the high economic growth from the early 1960s to the mid-1970s thwarted this policy. From 1965 to 1973 only 18 730 Africans in the Preference Area were resettled in the homelands.[13] After 1976 Africans in the Western Cape increasingly defied influx control, regardless of the penalties, because they found it impossible to survive in the poverty-stricken homelands. Even before the government abolished the pass laws and the CLPP in 1986, the application of the pass laws had effectively collapsed in the Western Cape.

The effect of pass laws and other labour controls on the different economic sectors will be analysed later. Here we briefly note its overall social impact.

First, and most important, was the alienating effect of the pass laws. Under NP rule prosecutions for pass 'offences' steadily rose from an average of 318 700 in the 1950s to an average of 541 500 in the years 1970 – 5. In effect, the pass laws reduced all Africans to the status of 'incipient criminals'.[14] Officials and policemen had the right to question Africans on the street and in their houses or places of employment. This meant that all Africans were harassed, even those who qualified to be in the cities.[15]

In the second place, influx control contributed to the distorted residential pattern of Africans. During the 1960s there was no net inflow of Africans to the white cities. By the early 1980s, South Africa, relative to countries on a similar level of development, was approximately 10 per cent under-urbanized. It has been estimated that in the 1960s and 1970s, influx control kept between 1,5 and 3 million Africans out of the white urban areas. Urban Africans faced a housing shortage of between 300 000 and 500 000 units. As a result of influx control and the ideological redrawing of political boundaries, an abnormally high proportion of Africans lived in the homelands — this figure rose from 39,1 per cent in 1960 to 52,7 per cent in 1980.[16]

Thirdly, employment patterns were also distorted. Despite the advanced stage of South Africa's economic development an unusually high proportion of the African work-force remained migrants. In 1985 Francis Wilson described the phenomenon in acute terms.

> In no other economy in the world has so large a proportion of the labour force stayed in single-sex accommodation and worked at a place where their families were unable to stay with them. Nor has any other modern industrial society seen such a process of oscillating migration continue for so long.... Exact figures are hard to come by but it is estimated that outside agriculture, perhaps two in every five black (i.e. African) men legally at work in the mines, factories and offices of South Africa are migrants.[17]

Finally the influx control contributed to a steady capital-intensification of industry because employers grew tired of all the problems related to the employment of African labour. In a country with an abundant labour supply, machines replaced men at a rapid rate. From the early 1970s a huge oversupply of African labour became apparent. Officials maintained there was only one solution for the army of unemployed: repatriation to their respective homelands.

Labour controls in mining and farming

From the perspective of the incoming NP government, the mines with their compounds and migrant labour-force approximated the apartheid ideal: Africans living out of sight and making no claim to citizenship and other political rights. This system continued until the early 1970s since the economic fundamentals of the gold-mining industry remained unchanged. Much of the ore that was mined was of a low grade; labour costs, due to inflated white wages, were high, and until 1970 the gold price remained fixed at 35 US dollars per ounce. Given the strong political power of the white miners, the mining houses lowered labour costs by cutting the black labour bill. Even then, profits were not stupendous. The average return on South African mining shares was 4,3 per cent for the period 1935 to 1963, compared to 7 per cent for United Kingdom equities.[18] It was only from the second half of the 1960s, but more particularly from 1970, when the gold price started its dizzy climb, that the mining houses were in the position to raise black wages substantially.

By contrast, the Chamber of Mines, which represented the main mining houses, had no stomach for driving down white wages. In fact, it had made possible the expression of organized white labour power on the mines. In 1937, following an agitation led by the Nationalist organizer, Albert Hertzog, the Chamber decided to recognize white trade unions and insisted that every white employee become a member of one.

The effect of this soon made itself felt. In 1953, an article in the *Mining Journal* remarked that pressure from white trade unions had led to an extension of the range of occupations from which Africans were excluded. In the 1950s and 1960s, the unions, headed by the powerful Mine Workers' Union, reinforced the colour bar by defining which jobs could not be done by Africans. The unions also guarded against any attempts by the industry to erode the colour bar through job fragmentation and alternative techniques of production. As a result of its tough bargaining, the average real cash wage of white miners nearly doubled between 1948 and 1973 while the ratio of white to African workers fell to its lowest ever. In fact, in 1953 it was half of what it had been at the time of the 1922 strike. The NP government supported and tightened the job colour bar on the mines. The Government Mining Engineer frequently fined mines for transgressions, most often for allowing Africans to use explosives without blasting certificates — a jealously guarded white preserve.

African labour remained unorganized, weak and poorly paid. The war boom had driven up the wages of Africans working in industry, but the mines managed to circumvent this trend by recruiting Africans from the territories beyond the northern borders, who were prepared to accept lower wages than South Africans. Under NP rule, foreigners as a proportion of the African work-force, jumped from 59 per cent in 1950 to 79 per cent in 1973. These foreigners enabled the mines to reduce their African wage bill. The average real cash wage of Africans stayed static between 1931 and 1962 and thereafter rose only slowly until 1971. The ratio of white to African wages widened from 1 : 12 in 1941 to 1 : 21 in 1971.[19]

The Chamber of Mines supported the migrant labour system and its underlying assumption that workers need not be paid to meet the subsistence needs of their families living in the reserves or in colonies north of South Africa's borders. In 1943 the Chamber of Mines declared to the Lansdowne Commission:

> The ability of mines to maintain their native labour force from the Reserves at rates of pay which are adequate for this migratory class of native but inadequate for the detribalized native is a fundamental factor in the economy of the gold mining industry.[20]

In its 1947 testimony to the Fagan Commission, the Chamber maintained that migrant labour was a stabilizing factor in the national economy and the national social structure. In 1960, when secondary industry began to demand stabilized African labour, the Chamber continued to defend the virtues of compound life, the levels of pay and the link with the African rural areas. It stated: 'The preservation of links with the tribal background imparts both security to the individual worker and stability to the economy of his home territory'.[21] As recently as 1979, Gavin Relly, who later became chief executive of Anglo American Corporation, declared: 'Migrant labour is here to stay... and should be accepted as a permanent feature of our economic and social order'.[22]

The apartheid state thwarted any attempts by more enlightened employers to change the migrant labour system. The opening of the Orange Free State gold-fields immediately after the Second World War could have been a turning point, but the incoming NP government resisted it. Concerned about an acute, skilled labour shortage and the wastage brought about by the high turnover of migrant labour, Anglo American Corporation announced in 1947 that it intended to stabilize 10 per cent of its African miners, and accommodate them in family houses in 'native villages'. Verwoerd, as Minister for Native Affairs, was resolutely opposed to this. He told Parliament in 1952 that this was an 'unhealthy development'. He believed that migrant labour served the interests of Africans and was preferred by them. Verwoerd would not agree to 3 per cent of the African labour-force on the gold-mines being stabilized. By 1970 the mines had stabilized less than 1 per cent of their African work-force. This was partly the result of close supervision by the apartheid bureaucracy which demanded that each case be justified.[23]

It is evident that on the gold-mines the apartheid state worked hand in glove with the white miners, using the long-standing system of migrant labour and the job colour bar to maintain a world in which whites prospered and Africans were poor and out of sight. In maintaining close ties with white labour, the government allowed the white miners to hold their employers to ransom over the job colour bar. This bargaining power under apartheid is best demonstrated by a 1967 agreement in which 22 000 union members won an 11 per cent increase in earnings conditional upon improved 'productivity'. However, as Francis Wilson notes, 'The increases in productivity were to come, not as one might expect from harder work by those benefiting from the agreement, but from the fact that blacks were to be allowed to do jobs that they had not been allowed to do legally before'.[24] Thus white workers received the lion's share of any increase in productivity resulting from the elimination of restrictive practices.

The dramatic rise in the gold price in the 1970s, from 35 to 800 US dollars, spelt the end of white labour's stranglehold. Gold mining became a hugely profitable and modernizing industry which was no longer willing to put up with artificial shortages of skilled labour. Addressing the Wiehahn Commission in 1978, the Chamber of Mines demanded that conventional and statutory restrictions be scrapped. In 1979 the white Mine Workers' Union attempted to stage 'another 1922' against the Wiehahn reforms, but the strike at the Okiep mine failed dismally. This was a turning point and Africans began to move steadily up the employment ladder. White capital had finally succeeded, after a century, in thwarting the job colour bar on the mines.

The large proportion of foreign labour on the mines, which was such a striking feature of the classic apartheid era, was greatly reduced in the 1970s. Owing to the disruption of the supply of foreign labour, African miners from South Africa (the boundaries as they existed from 1910 to 1976) were given preference in the queue for jobs, and their numbers rose from 21 per cent in 1973 to 57 per cent in 1982. This move severely hit Mozambique and Lesotho but there were indications that the citizens of the 'independent' homelands were also affected.

We turn now to farm labour. The Nationalist government favoured white farming interests and privilege, and engaged in large scale state intervention to keep the white farmer on the land and the rural areas white.

Under classic apartheid, the agricultural policy was aimed primarily at providing sufficient labour to the farms. During the Second World War the Smuts government had been unwilling to use compulsion to meet the farmers' labour shortages; instead it suggested that farmers improve conditions to entice workers to settle on farms rather than in urban areas. By contrast, the Nationalist government in 1951 answered the farmers' prayers by tightening the system of influx control and establishing labour bureaux. Before any African could leave a rural district, the local bureau had to be satisfied that the labour system in that district was adequate. In practice any African who worked on a farm found it virtually impossible to leave the farm legally for the urban areas.

As a result, a cheap, often brutally treated, African work-force remained trapped on the farms. In 1968 a deputy minister outlined the effect of influx control on African farm labourers as follows:

> A record of every registered Bantu farm labourer is kept in a central register, and the position is that the labourer cannot be employed in the urban areas, because as soon as

his service contract must be registered, it will be established that he is a farm labourer, and then he cannot be legally employed.[25]

In 1952 the average African farm wage was only 64 per cent of the cash wage on the mines and only 27 per cent of the manufacturing wage. African wages remained static until the mid-1960s, when increased mechanization and the need for a smaller, better trained labour-force began to push up wages. But by 1976 the average farm wage was still only a third of the manufacturing wage.[26]

Propping up white agriculture was an essential feature of the Nationalist project to keep South Africa a white land. During the 1950s and 1960s, the government introduced an elaborate system of state funding of white agriculture. In terms of this, favourable prices were set by the marketing boards for the farmers' produce, state aid was made available for price stabilization and subsidies, and there was massive funding of agricultural research. It has been estimated that, by 1970, state aid provided one-fifth of an average white farmer's income.[27] Although the NP government spent more on African farming than its predecessors had done, the disparity was still enormous. As late as 1973 it was still spending 'twice as much on the 90 000 rich, developed farmers, who had decades of favourable treatment behind them, as it spent on over half a million poor African farmers'.[28] This destroyed any chance that may have existed of homeland agriculture providing a viable alternative to migrant labour.

Finally, the apartheid order tried hard to stem the *verswarting* (blackening) of the white rural areas. A decade after the NP came to power, a commission found that 25 per cent of Free State farms were occupied solely by Africans, without any white supervision or presence, while in four magisterial districts in northern Natal, 42 per cent of the farms were operated in such a fashion.[29]

The government embarked on far-reaching measures to ensure that the rural areas remained white. It forced farmers in northern Natal to give up 'feudal' practices such as the labour tenant system. African tenants, who had lived on white farms for generations, were forced to take up wage labour or face the grim prospect of eviction to resettlement camps in the homelands.[30] As a result, the number of labour tenants fell from 163 000 in 1964 to 27 585 in 1970. Of a much greater magnitude was the resettlement in the homelands of farm labourers (particularly from the western Transvaal and Free State) who had become redundant.[31] As a result of large scale mechanization on the farms the wheel had turned: while the farmers had clamoured for African labourers in the 1950s, by the 1970s they wanted to get rid of 'superfluous' Africans on their farms. Some three-quarters of a million Africans left the farms between 1960 and 1975. By the end of the 1970s, however, the homelands were so overcrowded that officials sometimes asked farmers to keep Africans on the farms until they could be located somewhere else.[32]

A particular target of the policy of forced removals were Africans living in 'black spots'. These were areas of land acquired by Africans, usually prior to the Land Act, in the midst of white farms. Sometimes egged on by white farmers, the state removed people from these 'black spots' and dumped them in resettlement camps in the homelands. Usually these camps became too overcrowded for any agriculture and were often too far for commuting to work. Some scholars estimate the number of Africans removed from farms, squatter camps and black spots at approximately two million.[33] As Lipton notes, this removal of people from the land was not peculiar to South Africa; what was special about the apartheid order was the fact that the state forced people to move to overpopulated and impoverished homelands and away from the better opportunities available in the white towns.[34]

Agrarian apartheid slowly started to recede during the 1970s. Government policy shifted from favouring farmers to ensuring reasonable food prices for the cities. As the state became immersed in a fiscal crisis it cut back on farmers' subsidies. It also drew back from the forced removal of rural Africans. In 1984 President Botha acknowledged that 'the free Western world was extremely sensitive to large-scale removals of people'.[35] Comprehensive state intervention in the white rural areas lessened. South Africa's traumatic transition to capitalist agriculture was nearly over.

The industrial sector

When the NP came to power in 1948 it was concerned not only about the 'blackening' of the factory floor but also about the erosion of the traditional division between skilled whites and unskilled blacks. During the 1940s employers had overcome the skilled labour shortage by fragmenting skilled jobs and employing blacks as semi-skilled operatives. Increasingly the middle levels of the labour market had turned into a contested area for white and black workers.

The Nationalist government started out by applying the 'civilized labour policy' more strictly in the state sector, dismissing Africans and appointing whites in their places. Its first major legislation in this field was the Native Building Workers Act of 1951 which provided for the training of African building workers but prohibited them from working outside of African areas. This, in effect, extended the statutory colour bar to secondary industry; before this it had existed only in the mines. In 1956 the government went much further. By passing the Industrial Conciliation Act, it introduced what became known as job reservation. The Minister of Labour was empowered to reserve jobs for a particular race group or fix the ratio of European to African employees in an industry, a factory, or in certain types of employment. 'Work reservation is an attempt to create order out of chaos and to prevent racial friction', a government publication declared.[36]

The government mostly used its powers to protect the interests of whites. In 1958, for instance, the positions of ambulance drivers and firemen, and most ranks in the traffic police were reserved for whites, and in 1959 even the job of lift operator was reserved for whites.

The 1960s brought new challenges to the apartheid order. By this time virtually all white males were employed. They started to move out of manufacturing and into the white-collar jobs in the tertiary sector. Lack of competition among white workers led to slackness on the job, a high staff turnover and absenteeism. This constituted a serious bottle-neck for a manufacturing sector that was rapidly expanding. Increased mechanization created a large demand for machine operators, mechanics and technicians.[37]

To meet this demand, the government allowed some skilled jobs to be fragmented into semi-skilled operations in which blacks were employed. Informally blacks also did skilled jobs but were not given recognition or paid as skilled workers. White workers moved up to more senior jobs, usually in supervisory positions. This floating job bar introduced more flexibility which, in the 1960s and 1970s, was particularly beneficial to coloured and Indian workers. Nevertheless, the government remained determined to keep Africans out of skilled jobs. The 1970 Bantu Laws Amendment Act enabled government to prohibit the employment of Africans in any job, in any area, or in the service of any employer. This legislation was never applied and the tide of statutory apartheid in the job market began to recede.

In the end, formal job reservation by the government affected a fairly small proportion of jobs. Much more pervasive was the effect of the more informal job colour bar. This rested on negotiated agreements between employers and trade unions, and on custom. These informal arrangements worked hand in hand with broad rules governing the employment of blacks as laid down by government speakers. In the early 1970s, for instance, Africans were permitted to enter the lower grades in the steel and engineering industry, provided employers adhered to the following conditions: first, no person eligible for union membership — in other words, a white — was displaced; no African supervised the work of a white; and in the event of a recession, preference in retention or hiring was given to white workers.[38]

In contrast to the relatively well-researched manufacturing sector little is known about working conditions in the service sector under apartheid. One section that deserves particular attention is that of the large domestic service sector. In a sense, domestic service, and the concomitant continuous close contact between whites and blacks, violated the apartheid ethos. However, despite government attempts to keep blacks out of the 'white' cities at night, employers refused to give up domestics who lived on the premises. On the other hand domestic service also saw some of the worst forms of labour exploitation. Working hours were long and wages extremely low. The oppressive Masters and Servants Laws remained on the statute book until 1974. These laws, which made breach of service illegal, were applied almost exclusively to Africans in domestic and agricultural service. However, even in these sectors the law became obsolete. Prosecutions dropped from 43 000 in 1928 to 23 000 in 1968. When the law was finally repealed in 1974 due to overseas pressure there was hardly any outcry. It was generally recognized that the law benefited only the most backward employers and that more efficient laws existed to control labour.[39] But while other labourers were rapidly becoming unionized in the 1970s domestic servants remained unorganized and powerless.

The maze of laws and informal arrangements which were erected in the course of the twentieth century made it very difficult for blacks to organize themselves in trade unions. The Smuts government contemplated legislation which would give African workers, as far as practicable, the same rights as whites without permitting them to take part in industrial councils formed by employers and white workers. The incoming Nationalist government, however, objected to trade unions for Africans and prohibited strikes by African workers. It also closed off all formal channels for bargaining, barred the attendance of any African representative at the proceedings of an industrial council, and prohibited employers from facilitating the payment of union dues by African employees. The position of coloureds and Indians was somewhat better, but most of them belonged to segregated unions. These unions generally lacked the power to assert themselves and were often dominated by the white union. The situation only began to change at the end of the 1960s. Wild-cat strikes by black workers and the lack of negotiating structures for settling industrial disputes prompted employers to demand a reform of the labour laws.[40]

Education and training

Of all the factors constraining black advancement in the labour market, deficient education stands out. Often legal restrictions were of lesser significance than poor education which did not equip blacks to compete with whites in the labour market and made employers reluctant to train them properly. The apartheid order reaffirmed the long-standing tendency of the South African state to educate and train whites for the high and middle level job opportunities, and to provide the black groups with only a rudimentary education to do the remaining jobs. The small number of blacks who succeeded in reaching matric was considered sufficient to meet the needs of the teaching profession and selected categories within the civil service.

Occasionally Verwoerd's Bantu Education Act (1953) is projected as heralding a radical deterioration in educational standards for Africans. This is not quite correct for these standards were already very low when the NP came to power. By 1952 only 3 per cent of Africans had received more than a very elementary education. Among an African population of 11 million, only 8 488 had passed Standard ten — this was only one fifth of the number of whites who passed matric each year during the 1950s. By 1953, only 1 064 Africans had graduated from university — this was only twice the number of students who graduated at the University of Stellenbosch (one of eight white residential universities) each year during the early 1950s.

Nevertheless, in some areas the Verwoerd era represented a change for the worse. The introduction of 'Bantu education' killed the spirit of those church schools that did offer quality education. In 1953 there were 82 colleges for technical and professional education which whites could attend and 54 for blacks. By 1954, after Verwoerd's law came into operation, there were 88 for whites and only 21 for blacks.[41] Spending on education continued to favour whites disproportionately (see p. 106).

Communal Apartheid

Apartheid is often thought to be based on a white fear of being swamped or on an obsession with race purity. The NP's real motives were rather different. Racism and fear were significant only in conjunction with the Afrikaner nationalist claim to dominance. The survival of whites, and of the Afrikaners in particular, had to be safeguarded and the NP believed this could only be done through the policy of apartheid. Two considerations were inextricably linked: without a position of entrenched privilege the Afrikaners and the larger white community could not survive; without a separate white group a position of entrenched privilege could not be maintained.

In 1949 J. G. Strijdom, who went on to become Prime Minister, explained the connection in the following terms: 'If the white loses his consciousness of colour he cannot remain white'. He went on:

> The white population of our country, which is in the minority, can remain white only if they retain their consciousness of colour... [and] their national pride, their pride as a race... A sense of colour cannot be maintained on the basis of equality, that is, if there is no apartheid for daily intercourse in social affairs, politically, or in any other field.[42]

This was the underlying philosophy of communal apartheid.

The basic premise of communal apartheid was that white domination had to be safeguarded by statutorily classifying the race groups and banning sexual intercourse between whites and members of the black groups. This had to be reinforced by segregated residential areas and public amenities and by an education system which would make the black groups identify with their own communities. Such a communal basis was seen as a prerequisite for eliminating competition and conflict between whites and blacks in a common system.

Unscrambling the racial omelette

Drawing the lines and classifying people were exercises in unscrambling a human omelette. It was difficult enough when dealing with the Africans and the Indians who generally maintained distinctive cultural identities. But it was virtually impossible to classify, for the first time in history, coloured people and whites into statutory categories without inflicting untold human suffering. 'Don't let us trifle with this thing', General J. C. Smuts pleaded in Parliament in 1950, 'for we are touching on things which go pretty deep in this land'. He reminded the National Party that fifteen years earlier a Select Committee had found that a population register was impracticable and said that this was still true. In his view it was an attempt to 'classify what is unclassifiable'.[43]

The first of the Acts which imposed apartheid's statutory divisions was the Prohibition of Mixed Marriages Act of 1949 which forbade all marriages between whites and blacks. Between 1930 and 1950 the number of mixed marriages actually declined from 9,5 per 1000 marriages to 2,8 per 1000, and between 1943 and 1946 fewer than 100 mixed marriages per year took place against an annual total of almost 30 000 marriages of white couples. There was thus no immediate need for such a law. However, for the apartheid grand scheme it was symbolically important to curb all sexual intercourse across racial lines. In 1949 the NP passed a law which decreed that any union entered into in contravention of this Act would be declared 'void and of no effect'.[44]

The Immorality Acts of 1950 and 1957 were designed to curb extra-marital sex. This built on a 1927 Act which had prohibited all extra-marital sexual intercourse between whites and Africans. The 1950 Act extended this prohibition to all blacks. The 1957 law outlawed intimacy between white and blacks even if it fell short of sexual intercourse.

Before the Immorality Act was repealed in 1985 more than 11 500 people had been convicted under this Act and more than twice this number had been charged. Police used binoculars, tape recorders, cameras and two-way radios to trap offenders. Sometimes they examined the private parts of couples or took them to district surgeons for examination.[45] Some offenders committed suicide, others emigrated. Those who stayed faced wide-spread ostracism. In 1959 a NP newspaper acknowledged that the act was a 'harsh law, harsher in its impact than we imagined, causing nameless misery'.[46]

Nevertheless, the government retained the Immorality Act as it was considered indispensable to the political system. In a parliamentary debate on this issue in 1971, a NP speaker cited with approval an article by a law professor, S. A. S. Strauss, who proposed modifications of the law but not its abolition. According to Strauss, it regulated an issue which affected the whole political and social structure of the land.[47]

Another corner-stone of communal apartheid was the Population Registration Act of 1950. The government originally claimed that the purpose of this was to set up a population register and to issue all citizens with an identity card. In fact, its overriding objective was to create statutorily the apartheid communities and thus provide the foundation for the entire apartheid structure. In 1950, D. F. Malan correctly called the national register the basis of the whole policy of apartheid.

The Act enabled the state to classify every citizen as 'a white person, a coloured person or a native'. It also allowed for subdivisions within the African and coloured people. In 1950 it created three sub-groups within the category of coloured people : the Indian, Chinese and Malay groups. Being very small, the latter two groups were effectively lumped together with the general coloured group. From the start Indians were treated as a distinct category and they soon became the effective fourth statutory group under apartheid.[48]

The NP government was undaunted by the task of 'classifying what is unclassifiable'. The law used the categories of appearance, social acceptance and descent, with descent overriding the other two factors. The Act explicitly states that the tests of appearance and social acceptance are subject to the test of descent where the natural parents are known and classified. Initially, third parties could write to the official responsible and object to a person's classification. This opened the way for snoopers, as Smuts had warned in 1950. A person who was aggrieved by his classification could lodge an objection with an administrative tribunal generally known as the race classification appeal board.[49]

In 1986 a British analyst wrote in a Unesco publication: 'It is a testament to the single-mindedness of the regime in the period after 1948 that it was able to classify into largely self-constructed racial types an entire population of over 30 million people.'[50] It was only through single-mindedness, which allowed for little or no compassion, that such a shaky system could work. Racial identity, after all, is a sociological rather than a biological concept. In South Africa, centuries of mixing, particularly in the Western Cape, had resulted in so many borderline cases that persons born into one category could easily pass into another. Hence, it was possible for an individual to be classed as coloured or African while his blood-relations were classed as white or coloured.[51]

In proposing the population register in 1950, T. E. Dönges argued that in a population of 12 million the classification of only about 10 000 would 'cause difficulties'. They were mostly people who 'in appearance are Europeans, but who are in reality coloureds'.[52] The others, presumably, knew their racial group and would not cross the line.

In reality race classification unleashed untold suffering, particularly as far as the borderline cases were concerned. People were questioned about their descent and, in extreme cases, their fingernails were examined, and combs were pulled through their hair (if the comb was halted by tough curls the person was thought to be a coloured rather than a white). The Minister also under-estimated the size of the marginal group. In 1956 the Minister of the Interior said that officials had already dealt with 18 469 cases in which people had objected to their classification. The Minister added that over 90 000 borderline cases had already been encountered.[53]

The *Survey of Race Relations* of 1956 cited some cases which revealed the Kafkaesque racial maze in which individuals were trapped. In one case, a person whom the state deemed coloured had moved into a house previously occupied by whites. A coloured school teacher had then informed on him and he was subsequently charged with the unlawful occupation of the property. It eventually turned out that the man had lived his entire life as a European, as had two of his brothers. His wife's employer accepted her as white. However, two of his children were dark-skinned and were sent to a coloured school to save them from embarrassment, while another was fair-skinned and attended a prominent white school. The *Survey* also cited the case of an old man, aged 81, who had been married for 25 years. When census officials classified him as African and his wife as coloured they contemplated divorce proceedings for she could not face having to live in an African location. His appeal was allowed by the race classification appeal board.[54]

The NP asserted that classification was merely a technical exercise without serious implications for those not classified white. In 1950, an Afrikaans newspaper, *Die Burger*, expressed exasperation when the Leader of the Opposition, during a parliamentary discussion of the bill, branded race classification as a stigma. It asked: 'Why would it be a tragedy and a humiliation for those who pass for coloureds and who associate and live with coloureds to be classified as coloureds? Surely that's what they are (*Dit is hulle mos*)'.[55] D. F. Malan asked rhetorically in Parliament: 'Is it a stigma to call a Coloured man a Coloured man?'[56]

It was left to Judge van Winsen to put the matter in perspective:

> The decision as to a person's classification is, under the laws of this country, of cardinal importance to him since it affects his status in practically all fields of life, social, economic, and political. An incorrect classification can in all of those fields of life have a devastating effect upon the life of the person concerned.[57]

In 1974 a newspaper article captured the depth of suffering the Act could cause.

> The suicide of a 20 year old Coloured boy in Cape Town has brought to light an apartheid tragedy almost unparalleled in South Africa's history. The boy threw himself under a train at a suburban station when he learnt that his white girlfriend was pregnant. He could not marry her, because the Mixed Marriages prohibits marriage across the colour line.... The girl tried to commit suicide by cutting her wrists, but failed. She was unaware that her boyfriend was Coloured... The boy's identity is not being revealed, because his family is leading an illegal try-for-white

existence. The father, who is white, met the mother, who
is coloured, in 1950 — the year in which the Mixed Marriages Act was passed. They lived together, defying the
Immorality Act, which prohibits sexual relations across the
colour line. They have five children, and the family lived
as whites in a white suburb. They could not send the
children to school, because their birth certificates classified
them as coloured, and they would have been refused admission to a white school. This would have begun events
that would have led to their exposure, their expulsion from
the white suburb, and the loss of their white friends...[58]

Hand in hand with the racial sex laws and race classification went
the unscrambling of the residential pattern of South African cities. The
Group Areas Act was closely connected to the other laws regulating
communal apartheid. In the debate, A. J. van Rhijn, who subsequently
became a cabinet minister, explained the apartheid 'logic' during the
second reading debate: 'How can one maintain a law against mixed
marriages, how can one maintain a law against illicit intercourse
between white and black while people lived in mixed residential
areas?'[59]

The Group Areas Act did not have as devastating an impact on
Africans since they were already controlled under the Urban Areas Act
(originally the Native Affairs Act of 1920) except for a few townships
such as Alexandra in Johannesburg or Fingo Village in Grahamstown.
It was the two intermediate groups, the Indians and coloured people,
who were brutally hit by this Act. In the 1930s and 1940s residential
segregation had been increasing, but in Cape Town about a third to a
quarter of the coloured people were still living in suburbs integrated
with whites. Across the Cape Province white towns had *onderdorpe*,
literally 'lower towns', which also carried a class connotation. In these
areas coloured people lived interspersed with lower-income whites.
Before the NP came to power, coloured people had not been subject to
any control as to where they could acquire property. Up until the end
of the Second World War, Indians also acquired business premises and
houses in numerous towns in all the provinces, except in the Free State.

It was the large-scale penetration of Indians into 'white' Durban
which led to action by the Smuts government. In terms of a 1946 Act,
Indians could not buy or occupy fixed property in 'unexempted areas'
without government permission. In passing the Group Areas Act of
1950, the incoming NP government built upon the 1946 Act. It emphasized that this Act constituted a new direction in national policy that
would not only stop, but roll back integration all over the country.

Prime Minister Malan declared that the enactment of Group Areas will mean 'a fresh start for South Africa, because it heralds a new period'. He continued: 'What we have in this Bill before us is apartheid. It is the essence of the apartheid policy which is embodied in this Bill'. It was considered to be 'most crucial for determining the future of race relations'.[60]

In terms of the Group Areas Act the government acquired control over all inter-racial property transactions and inter-racial changes in occupation. It set up a Land Tenure Board (later Group Areas Board) consisting of white officials who could recommend the setting aside of particular group areas for the sole ownership or occupation or both of particular race groups.

Effectively this meant that no person belonging to a particular apartheid community was allowed to live in the 'wrong' group area. But the Group Areas Act went even further than this. It did not regulate only the ownership and occupation of domestic and business premises but also the provision of entertainment, in particular the 'occupation' of places of entertainment. This law effectively excluded blacks from restaurants, theatres, cinemas and sports clubs in white residential areas or in the central business districts.[61] As a rule this meant that in the white part of the town a black could only be served over the counter, and that it was virtually impossible for whites and blacks to have a meal or a drink together in a hotel, café or restaurant.

The Reservation of Separate Amenities Act of 1953 further entrenched public segregation. This Act was passed after the courts had declared invalid measures the Nationalist government had taken to reserve facilities on the railways exclusively for whites. Basing itself on the 'separate but equal' doctrine, the Appeal Court found that facilities for Africans were far inferior to those provided for whites. In the court's view the state did not have the power to provide blatantly inferior facilities.

The Nationalist government responded by passing the Separate Amenities Bill which was specifically aimed at legalizing unequal facilities for different races, thus obviating any court judgement that the degree of discrimination was unacceptable. For once the NP did not try to justify an apartheid law in terms of the benefits it would bring to the excluded. As P. W. Botha, then a back-bencher in Parliament, declared: 'If you stand for the domination and supremacy of the European then everything must in the first place be calculated to ensure that domination'.[62] The Separate Amenities Act provided the legislation required for the wide-ranging segregation of virtually every public facility, even to the point of absurdity. Until the mid-1970s the most salient feature of South Africa was 'separateness'; blacks were forced to use separate entrances to post offices, police stations and railway stations, separate buses and trains, separate parks, swimming pools and beaches, and separate public toilets.[63]

In justifying the Group Areas Act, the government used the inevitable friction argument. According to the NP, the Durban riots of 1949 (during which 142 Africans and Indians were killed in clashes between these groups) were a demonstration of ethnic tensions and the tragic consequences of residential integration. In introducing the Group Areas Act, Minister T. E. Dönges expressed the NP's apartheid philosophy succintly:

> It is the sacrifice we will have to make in order to bring about conditions most favourable for inter-racial harmony. For points of contact inevitably produce friction and friction generates heat which may lead to conflagration. It is our duty therefore to reduce these points of contact to the absolute minimum which public opinion is prepared to accept.[64]

Group Areas were also held up as promoting individual and ethnic self-fulfilment. The NP argued that there was nothing ethical in having coloured people live as 'appendages' in white areas where they would always have a feeling of inferiority. The apartheid alternative was sketched by D. F. Malan:

> [It] is only on the basis of apartheid in regard to residential areas that we shall be able to achieve sound relationships between the one race and the other. Only on that basis will we be able to secure justice for both sides. What justice is there for the non-European if he is in the position in which he is today? He will always have a sense of inferiority. He is unable to do justice to himself. On the basis of apartheid, however, with his own residential area, he will be in a position to do justice to himself. There he will be able to live his own life — there he can develop what is his own, and only by the maintenance and the development of what is your own can you uplift yourself and uplift your people.[65]

In proposing and implementing separate Group Areas the government constantly promised to provide increasing opportunities for blacks to start businesses, staff government offices and participate in local government within their separate townships. However, this form of compensation was extremely slow to materialize.

The NP also tried to minimize the effects of this massive exercise in social engineering. Dönges informed Parliament that 80 to 90 per cent of the available area in South Africa was predominantly occupied by persons of one group. The NP also claimed that the Act was not discriminatory since all groups would have to make sacrifices for separation. Margaret Ballinger was not impressed by these sentiments. In a 1950 speech she called them 'the greatest eyewash' and went on to say: 'We find it terribly difficult in this country... to implement the other side of the bargain'. Why? She replied that South Africa was a country 'where only white people sat in Parliament, and where practically only the white man had the vote'.[66] Even if one allows for the fact that rich people all over the world own much larger plots, Table 3.2 indicates the extent of the privilege enjoyed by whites who occupied five times more urban residential space per head than coloured people.

TABLE 3.2:   AREAS (IN HECTARES) OCCUPIED BY NON-AFRICANS UNDER GROUP AREAS ACT, 1987

	Whites	Coloured people	Indians
Total	750 070	101 797	515 005
Hectares per 1 000 of population	152,7	33,1	55,8

Source: Progressive Federal Party research division based on replies to questions in Parliament.

How the 'other side of the bargain' was kept is apparent from the figures for the removal of people under the Group Areas Act. By 1976, 306 000 coloured people (or 1 in 6 members of the group) and 153,000 Indians (1 in 4) had been removed, against 5,900 whites (1 in 666). In Cape Town the whole of the Table Mountain area to the west of the suburban line from Cape Town to Muizenberg was zoned for whites. The Cape Town City Council was so shocked by this move, and by the number of coloured homes and institutions that would be affected, that it boycotted a public hearing of the Group Areas Committee. Many well-placed observers believed that the government would find it impossible to implement its proposals.

The government confounded them. Over the following decades it moved coloured communities from areas such as District Six and Kalk Bay to sandy townships on the Cape Flats, or to the more attractive but distant suburb of Mitchell's Plain further east, or to Atlantis, forty kilometres to the north. Across the Cape Province coloured families were removed and usually settled further away from the 'white' town.[67] Although better housing was often provided and although the Act enabled thousands to become home-owners (instead of tenants in run-down town areas) great bitterness flowed from the fact that the removals were not voluntary. In many cases, extended families were broken up and the neighbourhood spirit destroyed, causing great sorrow and hardship. In the new townships a breakdown of social norms often occurred which gave rise to high crime rates. Many coloured home-owners suffered financial losses, while some white developers, who had brought property cheaply from the Group Areas Board, realized enormous returns on their capital.

In 1988, in one of the first joint parliamentary debates on a contentious issue, the Reverend Allan Hendrickse, leader of the Labour Party, told how his family had been dispossessed of the home and church built by his pastor father in Uitenhage. Addressing the NP representatives, he said: 'You can shake your head, but you stole my land, my people's land. This is the legislation of theft'.[68] Matters were seen in a completely different light by P. W. Botha who, as Minister of Community Development, directed the District Six removals. Hendrickse gave this account of his encounter with Botha.

> Last year, when I asked him to give District Six back to the coloured people, he wanted me instead to thank him for what he had done. He said he never heard of such ingratitude, that those people had been living in houses owned by whites, that they were exploited and living in slums, and that now they were home-owners in Mitchell's Plain. 'You must thank me for what I have done', he said, 'because now you are living in dignity'. That was his approach. It was typical of that gap in their thinking that just cannot be overcome.[69]

In the Indian community it was the traders who were hit particularly severely. By 1974 more than 5 058 Indian traders in the country had become disqualified in terms of the Act; out of these, 984 had at that stage been resettled. By 1977 the latter figure had increased to 1482. In Johannesburg the vast majority of Indian traders were evicted and only a few succeeded in re-establishing themselves as traders. In Pretoria, a third of the Indian traders in the Asiatic bazaars were allowed to remain, but even they suffered losses, for an important section of their clientele, the African and coloured populations living nearby, were forced to leave the area. In the Transvaal and Natal country towns Indian traders suffered devastating losses. In many towns the Indian community was small and traders had depended on trade with Africans and whites for 75 per cent or more of their turnover. Having to move out of the 'white town' to the Indian group area was often a lethal blow for traders, since they could not hope to make a living trading mainly with their own communities. In Durban, Indians retained their most important trading area, the Grey Street complex, but 12 000 Indian residents had to move.

Removals caused an upheaval in the Indian community. Large numbers in Durban had to move to the northern and southern fringes of the city, where they were further away from the city centre or their places of work than were whites. Moreover, extended families had to split for the first time and resettle according to individual financial means. As a result, the middle class (and traditional Indian leadership) moved to privately owned housing estates such as Verulam, while the lower-income groups moved into municipal housing in areas such as Chatsworth. Group areas produced a clear spatial expression of class differences in Indian life.[70]

As far as Africans were concerned, the government could, from the start, rely on the Natives (Urban Areas) Act of 1923 which enabled it to clear Africans out of the mixed residential areas which had sprung up in cities such as Johannesburg and rehouse them in locations. Needing additional powers, it passed the Natives Resettlement Act of 1954, in order to remove thousands of Africans, many of them home-owners, from Sophiatown and other parts of Johannesburg's 'Western Areas'. Sophiatown was rezoned for whites and callously renamed Triomf (Triumph).

Group Areas removals meant that houses had to be built on a large scale for those forced to leave. The first fifteen years of apartheid saw large scale housing construction for blacks. It was particularly noticeable in the case of Africans. There was, as Davenport puts it, 'a proliferation of small but well-built, if architecturally monotonous little boxes'.[71] In the five years ending in 1958, a total of 100 000 houses for Africans were built. As we have noted, the 1960s brought a near-freeze on housing construction in African townships in white areas. Instead the state concentrated on building houses in homeland towns located within commuting distance of work places. All this was part of the government policy to restrict the number of Africans living permanently in white areas to a minimum. Large housing shortages built up in the case of all three black groups. After insisting for years on formal housing, in the early 1980s the government finally started to allow organized squatting on the perimeters of cities such as Durban and Cape Town.

As Lemon points out, the removal of Africans to the outskirts has produced a distinctive feature of South Africa's urban landscape: 'the poor live at higher densities but further from the city than do whites', which is 'the reverse of the income gradient normally associated with Western European cities'.[72] The government has attempted to compensate for the adverse location of the black group areas by subsidizing public transport. However, many African workers, particularly casual workers, find even subsidized transport too expensive to afford. Many workers also have to spend long periods commuting, which lowers their productivity.

The NP did not want to turn the African group areas into self-sufficient communities, as they hoped to do in the case of the coloured people and Indians. By 1948 African consumers already formed one fifth of the domestic market. White interests, particularly those of the emerging Afrikaner traders and retailers, demanded that African wages be channelled back into the white economy. In 1958 the Afrikaanse Sakekamer opposed the granting of trading licences to Africans in the townships and, in 1959, the Minister of Native Affairs stressed that African traders were only temporarily resident in the locations. Once they had acquired the necessary capital and expertise, they were expected to move their businesses to the homelands.

In 1963 the government imposed further restrictions on African entrepreneurs. Africans in the 'white' areas were not allowed to run more than one business. They were denied the right to form companies or partnerships, barred from dealing in anything but day-to-day necessities, and prohibited from establishing African-controlled financial institutions and wholesale concerns. Municipal authorities were encouraged to persuade owners of existing dry-cleaning enterprises and garages to transfer their businesses to the homelands. A policy directive made it clear that only in the case of dire necessity would new trading licences be granted to Africans. In 1968 further restrictions were introduced. No African businessman was to be allowed to operate at more than one site and to sell goods to non-African persons who lived outside the urban African residential area. The government also introduced restrictions on the ownership of consulting rooms and offices by African doctors and other professional men in the townships.[73]

Segregated education had to provide the ideological cement for communal apartheid. When a Bill segregating university education was introduced in 1959, the Minister of Education stated the underlying philosophy:

> [The] Government's policy of separate development requires that non-Whites should be given every opportunity

to develop as individuals and for development as separate communities. If it is to be a balanced development, separate development demands that every individual national unit should produce from its own ranks, the necessary leaders, thinkers, educationalists, professional and technical people... [A] higher education can best be provided in one's own separate institution. Every national group of any consequence, if it wishes to hold its own, should have its own schools and its own university or universities — universities that not only serve as the focal point of its pride and self-esteem, but as a means to educate the community in the true meaning and value of university training as such.[74]

When the NP came to power in 1948, church and mission schools provided a large part of African education. The government forced most of these schools to close and in 1953 transferred the administration of African schools from the provincial authorities to the Department of Native Affairs which lay down departmental syllabuses and regulations. Despite the strong demand of urban African leaders for English-medium instruction, the government introduced 'mother tongue' instruction throughout African primary schools. Education was removed from general to communal departments. In 1963, legislation was passed which transferred control of coloured education from the provinces to the Department of Coloured Affairs. Similar legislation was passed for Indians in 1965.

The essence of apartheid education was to turn the child's point of identification away from the common society towards his/her own community. In 1959 the Minister of Bantu Education expressed the philosophy of apartheid education as follows:

> [We] must try to retain the child as a child of his own national community, because it is the basic principle of Bantu education in general that our aim is to keep the Bantu child a Bantu child... the Bantu must be so educated that they do not want to become imitators, that they will want to remain essentially Bantu.[75]

To a lesser extent the same was also true of coloured and Indian education. The government not only recognized the plural character of South African society, but presumed that it would continue indefinitely.[76]

Much attention has been given to the discriminatory government spending on the different population groups, the inferior quality of teachers, and the much larger ratio of teachers to pupils in black schools (see pp. 106 – 7). These have been major black grievances, but what was resented even more was an educational system designed to fit the apartheid model of society.

Political supremacy

Political supremacy operated at various levels. The Nationalist government simultaneously fostered and concealed Afrikaner rule. It set up the Afrikaners as the core group on whom the privileges and security of the larger white group depended. It thus recruited all whites, regardless of their party political affiliation, to uphold both the Afrikaner-controlled political order and the capitalist system. On another level, the government managed to divide the various black groups in order to rule them more effectively. Instead of the various black groups joining forces against the apartheid system, they were divided by the tensions it caused. Much of the political energy of blacks was consumed by battles between so-called 'radicals' and 'collaborators'.

Education for domination

Given the harsh disparities in wealth and the discrimination and suffering to which blacks were subjected, apartheid required a viable system of socialization and indoctrination. The system needed to convince whites, and Afrikaners in particular, not only that apartheid served their material interests as a group and as individuals, but that whites had a right to govern the country and were doing so justly, or at least as well as could be expected in a conflict-ridden divided society. The system thus required a set of beliefs that 'explained' the situation in a way that increased the solidarity of the ruling group. It also demanded that this group be shielded from the reality as experienced by black South Africans.[77]

In the task of constructing and disseminating this set of political beliefs throughout the educational system, the Afrikaner Broederbond played a crucial role. Started in 1918, the Broederbond, by 1945, had grown to a membership of 2 811, was organized in 183 cells and had assumed a leading role in the Afrikaner cultural, educational and economic mobilization. During the first decade of NP rule, the Broederbond lacked a sense of purpose. However, when Verwoerd became Prime Minister in 1958 he conferred on it the task of propagating a republic and promoting the new version of apartheid, called separate development. By the early 1960s the Broederbond had 8 000 members; this had risen to 12 000 by the late 1970s. At this stage Broederbond members occupied virtually all top positions in the political, civil service, church and educational hierarchies. All cabinet ministers were members, as was three-quarters of the NP caucus, virtually every senior civil servant, almost every principal of an Afrikaans university or college, half of all school principals and inspectors and 40 per cent of Dutch Reformed Church ministers.

Working closely with members of the cabinet, the Broederbond exercised constant vigilance to ensure that the education of Afrikaner children conformed to the teachings of the Afrikaans churches and to the history and culture of the Afrikaner *volk*, as defined by the Broederbond. This secret body also tried to impose a Christian National character on the education of all groups. Christian National education started from the premise that God decreed separate ethnic groups and that the prime objective of the school system should be to maintain these separate ethnic groups. Accordingly, education had to be rigidly divided along ethnic lines, with teachers of each group constantly inculcating a 'love for its own' (*liefde vir die eie*) among pupils. Although far less successful in the case of other ethnic groups, Christian Nationalism constantly reinforced the basic precepts of apartheid in white, and particularly Afrikaner, schools.[78]

School textbooks stressed racial or national differences among Africans. To various degrees they propagated the myth of racial superiority and of an epic Afrikaner struggle against British imperialism and the black majority, with the Battle of Blood River and the Anglo-Boer War as pivotal events.[79] In a recent study of school textbooks, J. M. du Preez found a pervasive use of 'master symbols', with the following heading the list: 'Whites are superior, blacks are inferior'; 'The Afrikaner has a special relationship with God', and 'South Africa rightfully belongs to the Afrikaner'.[80]

Given this educational background, it is hardly surprising that the attitude that Africans were basically different, if not inferior, was widespread in white society during the period of classic apartheid. In a study undertaken in the late 1960s, among white élites in Parliament, the civil service, and the business sector, 96 per cent of an Afrikaner sample agreed with the view that 'the Bantu should remain Bantu'; 83 per cent agreed that the 'Bantu was different by nature', and 69 per cent held the view that the 'Bantu was a child who was some centuries behind the white man in development'. The corresponding figures for a sample of English-speaking whites were 68 per cent, 73 per cent and 79 per cent.[81]

Politically aware Afrikaners could not be kept loyal to a system that rested on a policy of blatant racism, so the NP had to attempt to present apartheid as resting on an ethical basis. The first major effort in this direction was the Tomlinson Commission which, in a report submitted in 1956, propagated large-scale economic development of the homelands in order to turn them into viable political entities. Verwoerd rejected this proposal but reformulated apartheid as being a system which gave separate nations the opportunity to enjoy separate freedoms. Although the homelands policy of the 1960s and 1970s did not entail much more than administrative decentralization, Verwoerd's shift was remarkably successful in justifying apartheid to the overwhelming majority of Afrikaners. The long-term goal of total separation, however illusory, justified discrimination over the short to medium term.

It would nevertheless be a mistake to attribute the survival of the ruling group to a grim determination to cling to power and the apartheid ideology. Except for a brief period in the first half of the 1960s, the ruling group was never a closed corporation, immune to new ideas and values. The government, however reluctantly at times, listened to the representations of think-tanks, pressure groups and lobbies. The press remained free to warn of danger signals. Conferences continued to stress the need for marginal political and economic concessions to blacks. Initially Afrikaner businessmen and academics exerted most influence. In the 1970s, some English-speaking businessmen also began to make substantial inputs, and in the 1980s, the government started to seek out moderate black opinion and consulted attitude surveys before embarking on policy decisions.

At school level, however, most white youth remained blissfully unaware of the political reality. How far the worlds of Afrikaner and African children were apart was evident from a study of matriculants from these two groups, which was undertaken on the Witwatersrand in 1986. Nearly 80 per cent of the African sample believed that Africans should be allowed to stage public protests, against 12 per cent of the Afrikaners; 95 per cent of Africans believed in equal political rights for all, against 12 per cent of the Afrikaners.[82]

Black political subordination

Political supremacy was apartheid's objective right from the start. A NP pamphlet entitled *Apartheid and Guardianship*, published just before the 1948 election, clearly spelt out the nature of domination under apartheid: 'The right of the non-European to exist and to develop is acknowledged, but apart from and under the guidance of the European'.[83]

The establishment of political supremacy occurred in three phases.

First there was disenfranchisement and the attempts at denationalization, then subordinate institutions were set up and finally, in the era of reform-apartheid, joint bodies were created in which white power and interests remained nevertheless decisive.

During the first two decades of apartheid the government removed all forms of black representation in Parliament. Its immediate target was the coloured vote on the common roll. The election of 1948 was won by a narrow victory of only eight seats, and a minority of the popular vote. For the NP it was obviously intolerable that approximately 50 000 coloured voters could theoretically determine the result of the following election. Although this fear of defeat was quite powerful at first, it was not the decisive factor. After the 1953 election, the NP conceded that the coloured voters had not affected the results in a single seat. Despite this, the NP continued its assault in the face of a major constitutional crisis, until the coloured voters were finally removed from the common roll in 1956.

More important were two other interrelated reasons. Firstly, there was the NP's all-consuming drive to deflect the African demand for representation in Parliament. Secondly, there was the NP's growing insistence on consistent thinking along apartheid lines. It believed that it was basically unsound to have coloured people on a common roll, or to allow even a limited number of coloured leaders to be elected to Parliament on a separate roll. In the NP's view, this created a disastrous precedent since the African majority would constantly aspire to the same rights. Furthermore, with its penchant for internally consistent models, the NP believed that coloured representation constituted a fatal flaw in the apartheid system which was based on the principle that the various racial groups should develop separately.[84]

Initially the NP shrank from complete disenfranchisement and after coloured people were removed from the common voters' roll they were still able to elect four white representatives to Parliament and two in the Cape Provincial Council on a separate roll. However, in 1970, this representation was also terminated.

Similar thinking lay behind the NP's scrapping, soon after it came to power, of legislation which provided communal representation for the Indians in Parliament and the Natal Provincial Council. The same also applied to the NP's hostility to the Natives Representatives Council (NRC) which ceased to exist in 1952. As Verwoerd explained in 1948, the NRC could develop into a separate Parliament, creating a potential for conflict.[85]

In 1959, all African representation in Parliament (seven white MPs and senators) was abolished. Verwoerd argued that, as a quid pro quo eight different African nations would be allowed to develop separately in their own homelands, possibly up to the point of becoming independent. Urban Africans were to enjoy political representation only in their separate homelands.

In Verwoerd's view it was only logical to remove the African representatives from Parliament. He said in 1959:

> The Native Representatives cannot be retained because they are declared enemies of this policy of emancipation. What they say in Parliament is not so important, but it is because of the access they have to the Natives, and the status they have acquired to create the wrong impressions in the minds of the Bantu masses, to try to dissuade them from the process of growing independence, that they should be deprived of this opportunity to abuse their position.[86]

Disfranchisement was part of a general attempt to impress on blacks that they were not to consider themselves full citizens and part of the nation. Compelling blacks to attend separate universities was a small part of this attempt. A coloured educator, Richard van der Ross, saw the connection. When, in 1959, a cabinet minister said that blacks at an open university would try to attain to that which they could never be, Van der Ross wrote that this was a direct warning to coloured people that they should not try for full citizenship and, in particular, not insist on the 'fundamental right to pursue higher education freely'.[87]

Verwoerd was quite blunt about his view that coloured people were not part of the South African nation. Coloured people were not allowed to participate in the referendum on the question of South Africa becoming a republic in 1960. Verwoerd said a year later: 'Let me be very clear about this: when I talk of the nation, I talk of the white people of South Africa'.[88] The true destination of the coloured people, in Verwoerd's view, was to develop along parallel lines until they became a nation in their own right. Van der Ross wrote: 'It means that in future no non-white South African need regard *Die Stem* as his national anthem, or the South African national flag as his flag'.[89]

In effect, this was also Verwoerd's view. In 1965 he argued that as much as Ceylon was the land of the Ceylonese, regardless of the existence of the Tamils, South Africa was a white state despite the existence of other groups. After all, in Verwoerd's reasoning, the Bantu were being 'eliminated' from the political life of the state, and the coloured people and Indians were mere marginal minorities.[90] After Verwoerd's death, the Cape wing of the NP rejected all talk of a coloured homeland and included the coloured people in their definition of the nation. In 1972, using the distinction in Afrikaans between *volk* and *nasie*, P. W. Botha declared that coloured people were part of the nation, a term which refers to matters of state, but not part of the *volk*, which is an ethno-cultural concept.[91] This was the beginning of the road to the Tricameral Parliament which formally incorporated Indians and coloured people into the definition of the nation.

In the case of the Africans the government went much further along the road of denationalization. In terms of the Bantu Homelands Citizenship Act of 1970, all Africans were regarded as citizens of the different homelands. As we have seen, the Bantu Laws Amendment Act of 1978 decreed that children of citizens of independent homelands born after independence, would not acquire Section 10 exemptions, even if they had been born or resided in an urban area in 'white South Africa' (see p. 94).[92]

Given this background it is hardly surprising that, with very few exceptions, no black person held any senior position in the white central state apparatus or provincial administrations. This was partly due to the inferior education and training and the job colour bar, and partly to the apartheid policy which stated that better-trained black people should first and foremost serve 'their own people'.[93] Departments which dealt with the administration of the black communities, soon became the fiefdoms of bureaucrats strongly committed to apartheid.

For a long period virtually no blacks were appointed to statutory bodies. In 1950, a minister was asked whether he would allow any blacks on the Land Tenure Advisory Board which controlled the transfer of property from one group to the other. He replied: 'There is no provision against it, but I see no reason for allowing non-Europeans on the Board'.[94] This set the tone for the administration of apartheid.

Political apartheid was essentially about the establishment of subordinate structures for the disenfranchised blacks. The more idealistic NP supporters continued to plead for these structures to be made politically more meaningful. An editorial which appeared in *Die Burger* in September 1948 was typical of this sentiment: 'A policy of apartheid must not only take away but must give and give generously'.[95]

The apartheid order failed to make good on its promise to provide adequate alternative structures. The absence of any direct electoral pressure made the government complacent and insensitive to the feelings of blacks. For coloured people, the government was content to establish dummy bodies to act as intermediaries between the government and the community. These dummy bodies were allowed to stumble on despite their patent defects. The Union Council for Coloured Affairs, which was projected as part compensation for the removal of coloured voters from the common roll, lasted for ten years without making any impact. This nominated and consultative body was replaced in 1970 by the Coloured Persons Representative Council, which lasted another ten years. Held up as a proper substitute for the abolition of all coloured representation in Parliament, it was elected but had no real power. Voters were conscious of the council's impotence and the poll at elections dropped. In 1975 only 25,3 per cent of the potential electorate voted, against 35,7 per cent in 1969.

By the mid-1970s the coloured middle class, always strongly opposed to 'parallelism', was bitter and alienated. Not only were they disenfranchised, but their route upwards in the Department of Coloured Affairs, the University of Western Cape, and the various training and technical colleges was blocked by whites, many of whom were adherents of apartheid. Only by the late 1970s did the impact of 'coloured empowerment' become visible in these institutions.[96]

In 1970 the NP removed the coloured municipal franchise. No viable alternative existed since coloured townships were essentially dormitory towns without a sufficient revenue base to sustain separate municipalities. The coloured management committees, which the government established for these townships in the place of town councils, were largely toothless — all power was in the hands of white municipalities or the provincial administration or central government.

The same development occurred in the case of the Indian group. In 1961 the government established the Department of Indian Affairs to serve as a channel for communication. In 1968 this was complemented by the South African Indian Council, a purely consultative body. It was envisaged that functions relating to Indian administration could be delegated to this body once it became elected. This did not happen until 1984. Indians successfully staged boycotts of the Local Affairs Committees which were supposed to compensate for their lack of municipal franchise.

The gap between promise and delivery was even larger in the case of the Africans. Since the homelands were supposed to compensate for representation in the white state, one would have expected the rapid development of these areas as alternatives to political integration. In a discussion with Verwoerd in 1961, the Secretary General of the United Nations, Dag Hammerskjöld, described the requirements which had to be met if the government wanted the homelands to be considered a 'competitive alternative':

- there had to be sufficient and coherent territory for the homelands;
- there had to be rapid economic growth and industrial development;
- Africans working outside the homelands should return only on a voluntary basis and had to have their human rights recognized and
- the government needed to let the homelands proceed fairly rapidly to full independence.[97]

In the two decades following this conversation the government made it only too clear that it was not interested in the first three requirements. There was no attempt to increase the land allocated to the homelands over and above the 13 per cent which had been laid down in 1936, when the government had had far less ambitious plans for the homelands. Far from recognizing the human rights of urban Africans, the Nationalist government used the homelands as a justification for the 'repatriation' of surplus Africans and as an excuse to deny Africans in white areas citizenship and their South African nationality.

Between 1959, when Verwoerd made his momentous speech, and 1971, when the Bantu Homelands Constitution Act was passed, the economic and political development in the homelands was largely neglected. Verwoerd's prohibition of investment by white entrepreneurs in the homelands constituted a severe blow to growth. Between 1960 and 1972 a total of only 85 554 jobs were created in the homelands and border areas, well below the figure of 50 000 new jobs a year which the Tomlinson Commission had considered necessary.

The government's resettlement policy and strict curbs on African urbanization greatly impeded the prospects for growth. Apart from the rapid natural population increase, between 1960 and 1980 the homelands had to accommodate one million Africans removed from the farms, 600 000 from 'black spots' and three-quarters of a million removed from white cities under a policy of township relocation. For the homelands the effects were devastating.

Whereas the average population density in the homelands between 1918 and 1950 was 50 to 60 persons per square mile, this had doubled to 125 persons per square mile by 1970. Between 1970 and 1980 the overall population of the homelands rose by another 57 per cent. The result of the massive inflow of people was a dramatic decline in agricultural production.[98] The homelands became overwhelmingly dependent on the earnings from migrant labour. To give KwaZulu as an example. In 1960, 54 per cent of its Gross National Income was earned outside the homeland. This rose to 74 per cent in 1970 and to 80 per cent in 1976.[99]

In the first half of the 1970s the government stepped up its development efforts in the homelands. Between 1970 and 1976 the proportion of the South African budget spent on homelands rose from 6,3 to 8,1 per cent. Nevertheless the homeland governments remained almost totally dependent on Pretoria. KwaZulu, for instance, continued to rely on financial transfers from Pretoria for between 70 and 80 per cent of its income. In the case of independent homelands such as Transkei and Bophuthatswana such transfers represented 70 and 50 per cent of their respective incomes in the early 1980s.

Thus Pretoria risked little by giving self-government or even nominal independence to these political entities. It was well understood among homeland leaders that any show of opposition could lead to devastating punitive measures such as curtailment of financial assistance or the repatriation of migrant labourers. Moreover, the Bantu Authorities Act of 1953 had greatly strengthened the position of the chiefs on local and regional levels. They were by and large conservative elements, keen to preserve the status quo. Nevertheless, by 1972 only Transkei had been granted self-government.

From the early 1970s a visible middle class, comprised of politicians, civil servants, teachers and businessmen, emerged in the homelands. Although they are rather more progressive than the chiefs and are more critical of the South African government, they have the same stake in the existing political order.

Instead of the devolution of any real political power the homelands represent the devolution of administrative functions. By the early 1980s the South African government was spending approximately 9 per cent of its budget on the homelands, but about 60 per cent of the total current annual budget of the homelands was committed to regular expenditure, that is, education, health and industrial infrastructure. It could thus be said that initially the homelands met the South African government's political and ideological objectives without any prohibitive costs. However over the medium to long term, the expenses relating to the duplication of administrative structures in the homelands and in the common area tended to become a major burden (see p. 130).

Local government for Africans in white urban areas remained in the doldrums until the 1980s. As part of the drive to remove all African representation in the white state, Verwoerd abolished Location Advisory Boards, which had not been much more than consultative bodies. Surprisingly, in 1961 the government introduced Urban Bantu Councils which were elected; but these were powerless bodies which enjoyed little respect. From 1973 urban African administration was vested in bureaucratic bodies called Administration Boards. Apart from controlling African influx and registering migrant workers, they also dominated African local authorities. In 1977 the Urban Bantu Councils were replaced by Community Councils which also enjoyed little power or prestige. Even in its own terms, the government's attempt to compensate Africans for their exclusion from political representation in the white state fell far short.

In a sense it is correct to regard the draconian security legislation which the government passed in the 1950s and 1960s as a consequence of apartheid. However, had South Africa after 1948 pursued the Smuts government's hesitant road to integration rather than apartheid, the situation with respect to human rights may not have been significantly better. Unless either whites or blacks had completely capitulated, conflict — and curbs on civil liberties — would probably have occured in the South African situation, especially as blacks became better educated, more urbanized and increasingly impatient with the denial of their full civil rights.

Apartheid created serious divisions within black ranks. This fractured the black challenge to white supremacy. The most serious division has been that between the ANC, which went into exile in 1960, and Dr Mangosuthu Buthelezi who set up his Inkatha movement under the umbrella of the KwaZulu government.

The politics of white privilege and patronage

Apart from protecting whites in the private sector (see pp. 78 – 80), the apartheid order also spawned a whole set of policies which favoured whites over the black groups, and a particular section of the white group, namely the Afrikaners, over other whites. Partly as a result of these policies, the income distribution of whites and the black groups remained badly skewed (see Table 3.3).

First we shall briefly discuss the policies of patronage which benefited Afrikaners over other whites.[100] Apartheid formed a common platform on which all the various classes within Afrikanerdom were able to join forces for nearly three decades, with the common purpose of advancing Afrikaner interests. Since the government was committed to the policy of apartheid and to enforcing bilingualism in the public sector, there was a rapid Afrikanerization of the civil service. In 1948 the higher ranks in many departments were still predominantly occupied by the English-speaking supporters of the United Party; twenty years later Afrikaans-speaking Nationalists filled virtually all senior positions.

TABLE 3.3:   SHARE OF INCOME PER PERSON EMPLOYED ACCORDING TO
ETHNIC GROUP, 1946 – 76
(Size of each population group as a percentage of the total population is
given in parentheses).

Year	1946		1960		1976	
Whites	74,0	(21,05)	70,0	(18,56)	63,0	(16,82)
English	44,5	(8,77)	37,0	(7,78)	31,5	(7,04)
Afrikaners	29,5	(12,28)	33,0	(10,78)	31,5	(9,78)
Coloureds	4,6	(7,89)	5,5	(8,98)	7,6	(9,78)
Asians	1,8	(2,63)	2,0	(2,99)	2,7	(2,94)
Africans	19,6	(68,24)	22,5	(69,46)	27,0	(70,45)
Total	100,0	100,00	100,0	100,00	100,0	100,00

Source: Calculations by Sampie Terreblanche reproduced in Heribert Adam and
Hermann Giliomee, *Ethnic Power Mobilized: Can South Africa Change*? New
Haven: Yale University Press, 1979.

The Nationalist government also used the rapidly expanding public
or semi-state corporations to promote Afrikaner economic progress.
By 1968 there were twice as many Afrikaners in the state and the
semi-state sector than had been the case before 1948. The share of
Afrikaner businesses in the private sector also grew from 10 per cent
in 1948 to 21 per cent in 1975. This was partly as a result of favouritism
although this is often exaggerated in the popular media. Farmers, of
whom more than 80 per cent were Afrikaners, were greatly aided by
favourable prices set by the marketing boards and other forms of
government aid and intervention.

The rapidly shrinking gap in the incomes of Afrikaners and other
whites is shown in Table 3.4

TABLE 3.4:   RATIO OF INCOME OF AFRIKANERS TO ENGLISH-SPEAKERS,
1946 – 76

Year	Personal income	Per capita income
1946	40 : 60	100 : 211
1960	47 : 53	100 : 156
1976	50 : 50	100 : 141

Source: Adam and Giliomee, *Ethnic Power Mobilized: Can South Africa Change*?
New Haven: Yale University Press, 1979.

Apartheid also boosted the privileges that whites received in the
social services provided by government. In the decade before 1948 the
state had made some visible progress towards expanding the social
services and narrowing the wide gap in the social services provided to
whites and the various black groups. For instance, in 1944 old age and

blind pensions were extended for the first time to Africans and Asians (see also p. 23). Africans received approximately 30 per cent less than whites while Asians and coloured people were paid at half the white level.[101]

The first ten to fifteen years of NP rule saw a reversal of trends. The government blatantly insisted on the priority of white interests and the need to scale back the services provided to blacks. Bromberger captures the development well:

> The early years of the NP government ... were characterised by a direct and purposeful assault on aspects of the new fragile welfare construction that had begun to emerge during the previous decade. Several, though by no means all, of the recent advances were reversed. Unemployment insurance became less inclusive; black [i.e. African] school-feeding disappeared, black educational expenditure was tied once more to the blacks' capacity to pay.[102]

By 1960 the disparities in pensions between the four racial groups had increased: coloured people were paid 40,2 per cent, Indians 35,9 per cent and Africans 14,9 per cent of the amount received by whites. The 1960s saw the beginnings of a slow process of equalization. In the early 1970s the government pledged itself, over an unspecified period, to the elimination of the discrimination in wages of public employees and transfer-payments (such as pensions), and slowly began to make progress in this direction. In the case of pensions, in 1972 the coloured and Indians had regained the 1947 level of 50 per cent of the white level and by 1980 the maximum of both groups stood at 56,9 per cent of the white level. African pensions dropped to a low of 13,2 per cent of the white level in 1966, but this figure rose to 30,3 per cent in 1980.[103]

Spending on education continued to favour whites disproportionately. As Table 3.5 shows, spending on whites as a group vastly outstripped spending on the other groups — nearly five times as much was spent on the white group than on Africans in 1952 when the government pegged expenditure on African education to African taxation. This rose to seven times as much in 1960. During this period the expenditure per African pupil declined from R17,08 in 1954, to R13,08 in 1960 and down to R11,56 in 1963.

The gap in expenditure on education of the various groups slowly started to narrow from the early 1970s. The shortages of skilled workers experienced by employers led to a massive increase in the amount spent on the education of all of the black groups. Particularly prominent is the real increase of 206 per cent spent on African education between

1977 and 1982. In 1987 for the first time, a roughly similar amount was spent on African and white education. Spending on whites declined in real terms between 1982 and 1987. Roughly similar trends are observable in the case of the two smaller black groups. Table 3.5 gives a good indication of the rythmn of apartheid — the severity of the 1950s, particularly as far as Africans were concerned; the slow relative improvement in state spending on blacks in the 1960s; the substantial real increases in expenditure between 1970 and the present and, in particular, in the period between 1977 and 1982.

TABLE 3.5:   STATE SPENDING ON EDUCATION

	A	B	C	D	E	F	G	H
1952	874 582	n/a	99 706	n/a	27 319	n/a	144 385	n/a
1957	969 553	10,86	122 561	22,92	38 213	39,88	165 776	14,82
1962	1 280 105	32,03	146 742	19,73	49 960	30,74	169 532	2,27
1967	1 747 764	36,53	289 399	97,22	97 031	94,22	254 344	50,03
1972	2 719 104	55,58	357 346	23,48	152 092	56,75	476 671	87,41
1977	3 181 656	17,01	523 088	46,38	220 598	45,04	640 922	34,46
1982	4 098 822	28,83	807 884	54,45	390 698	77,11	1 959 922	205,80
1987	3 320 700	−18,98	1 007 569	23,97	404 647	3,57	3 400 250	73,49

A = White expenditure in real 1987 rands (R'000s)
B = % real increase
C = Coloured expenditure in real 1987 rands (R'000s)
D = % real increase
E = Indian expenditure in real 1987 rands (R'000s)
F = % real increase
G = African expenditure in real 1987 rands (R'000s)
H = % real increase

Source: Researched and compiled by Monica Bot.
Note: African figures include TBVC states.

The white share of the total South African income per person remained at approximately 70 per cent between 1946 and 1960, despite the fact that the white population as a proportion of the total population declined. The African share remained stagnant at about 20 per cent (see Table 3.3). Other calculations, such as those done by McGrath, show that the respective shares remained constant until 1970. The decade from 1970 – 80 saw some dramatic changes, due mainly to the government policy to end salary discrimination and the wage increases for blacks within the private sector. As a result the white share of the total income declined to 60 per cent while the African share rose to just below 30 per cent.[104]

The pattern of taxation and state spending enabled whites to justify the government's politics of patronage. In 1949 – 50 whites paid

81,1 per cent of the taxes; this figure declined slowly to 76,9 per cent in 1975 – 6, while the African share of taxes rose from 11,4 per cent to 16,2 per cent in the same period. Whites, as the major tax-payers, claimed that they were entitled to a major share of the social services. According to McGrath, in 1949 – 50 whites received 61 per cent of the state expenditure directly allocated by racial group, while the coloured, Asian and African shares were 11 per cent, 3 per cent and 25 per cent respectively. The proportions stayed fairly constant until 1969 – 70. By 1975 – 6 the white share had declined to 56 per cent, while that of Africans had risen to 28 per cent. The proportions for coloured persons and Asians stood at 12 per cent and 4 per cent respectively.[105] Blacks argued that, by spending more than half the social welfare budget on whites, the system perpetuated white supremacy.

Some redistribution by the state did take place. McGrath estimated that the per capita income of whites before redistribution was thirteen times that of the three black groups together, but after redistribution the white income had dropped to eight times that of the black groups. The gap between whites and blacks was still large — as one would expect in view of the disenfranchisement of blacks. However, as Bromberger remarks, the fact that there was any redistribution at all requires explanation, especially from those who argue that apartheid was devised exclusively to serve white interests.

Conclusion

Four months after the NP victory in 1948, a major debate took place in Parliament about what the new government's policy of apartheid entailed. Some speakers argued that it was merely a modification of the long-standing segregation policy. However, General Smuts told the House that he had been informed that the new policy was 'something quite unique'.[106]

Although it was built on segregation, apartheid tended to reaffirm its uniqueness rather than its continuity with its antecedents. In the period prior to 1948 there had been nothing comparable to the forced removals which occurred between the early 1960s and the early 1980s. Indeed, nothing symbolizes the oppressive nature of apartheid more than the massive scale on which population relocation occurred.[107]

When one looks at the elaboration of the system of migrant labour within a world context, apartheid's uniqueness stands out. Nowhere else has migrant labour dominated a country's labour system to the same extent. Perhaps no large labour-force in the modern era of world history was quite as rightless as South Africa's migrants were during the 1960s and 1970s. The regulations to which they were subjected prompted John Rex in the early 1970s to state that South Africa had possibly developed one of the most perfect systems of labour control in the world. In Rex's words, laws enable 'the employer to control a variety of forms of legitimate violence which he may use against his workers, but at the same time do not require that he should buy the worker for life and be responsible for the worker for life'.[108] It should be stated here that this system of labour control evolved and was not consciously designed in all its aspects.

At the same time, the degree to which white workers were cushioned and protected between 1948 and the early 1970s, was probably also unique. Only the proportional demographic decline of the white group after 1960, together with the rapid economic growth of the 1960s and early 1970s (see pp. 116 – 18) managed to break white labour's stranglehold on the job market.

The policy of communal apartheid, with its drastic race classification, racial sex laws, and urban segregation was one of the most radical exercises in social engineering in the world. In the cities each group was supposed to develop its own communal life. In their systematic planning the 'apartheid cities' were recognizably different from the 'segregation cities' which had preceded them.[109] Communal apartheid ensured that intergroup contact was kept to a minimum. It also provided the basis of segregated education, health facilities and social services.[110] The great majority of black townships remained soulless dormitory towns; the inhabitants spent their money in white towns and their civic affairs were controlled by the local and central white state.

But even this did not constitute the unkindest cut of apartheid. It can be said that modern man's greatest desire is to be regarded as a unique individual rather than a category. The apartheid order treated blacks as categories; moreover, it pretended to know better than they themselves what was good for them. 'God made me a man; you made me a coloured man', Allan Hendrickse bitterly told the NP in a joint sitting of Parliament in 1988. It was reported that the speech made a great impact on NP parliamentarians. This is not surprising. As the Afrikaans poet Breyten Breytenbach wrote:

> The first thing to point out is that apartheid works... [It] has effectively managed to isolate the white man. He is becoming conditioned by his lack of contact with the South Africa inside himself... His windows are painted white to keep the night in.[111]

As regards political disenfranchisement, the apartheid order is not unique. Genuine democracy is rare in the modern world. Turning natives into strangers in their own land has also happened in places such as Israel and Northern Ireland, and major income inequalities between groups are found the world over. However, in South Africa this takes an extreme form. It is rare for one group to control more than 90 per cent of the instruments of production, in other words, the capital and land. That this group is racially defined makes it worse.

Apartheid is also exceptional for a different reason. The homelands comprise 25 per cent of the arable land in South Africa. There are few countries in which the poor hold a fairly sizeable part of the arable land. Apartheid's recognition, albeit half-heartedly, that other ethnic groups are in principle equal to whites, has eventually made blatant discrimination with respect to salaries and social services for blacks untenable. Although often done grudgingly, a slow process of equalization did start in the course of the 1960s — which is remarkable given the fact that blacks were disenfranchised. This process accelerated from the mid-1970s, and now whites generally accept the need for equal salaries within the civil service.

The implementation of apartheid in the 1950s and 1960s was characterized by a great deal of fervour and a belief in the intrinsic value of the policy. By 1970, however, the policy had acquired a predominantly instrumental character. Government spokesmen began to present apartheid as the most functional way of ordering society. It was this shift from a missionary sense of purpose to a pragmatic functionalism that made significant changes in the labour laws and communal apartheid possible. In the next chapter we shall discuss these changes in the evolved form of apartheid which we shall call reform-apartheid.

Notes

1. For an excellent critique of this view, with special reference to influx control, see Deborah Posel, 'Influx Control and the Construction of Apartheid', doctoral dissertation, Oxford University, 1987, esp. pp. 7 – 17, 339 – 44.
2. Piet Cillié, 'Bestek van apartheid : Wat is (was) apartheid?' *Die Suid-Afrikaan*, Spring 1988, p. 18.
3. *Die Burger*, 18 May 1968.
4. For a comprehensive analysis of the system of labour control see Stanley Greenberg, *Legitimating the Illegitimate: State, Markets and Resistance in South Africa*. Berkeley: University of California Press, 1987.
5. Cited by Merle Lipton, *Capitalism and Apartheid : South Africa, 1910 – 1986*. Aldershot: Wildwood House, 1986, p. 18.
6. The following discussion of the evolution of the influx control system draws heavily on Posel, 'Influx control and the construction of apartheid'.
7. Douglas Brown, *Against the World : A Study of White South African Attitudes*. London: Collins, 1966, p. 114.

8. Lipton, *Capitalism and Apartheid*, p. 35.

9. Lipton, *Capitalism and Apartheid*, p. 153.

10. Keith Gottschalk, 'Industrial decentralization, jobs and wages', *South African Labour Bulletin*, vol. 3, no. 5, p. 51.

11. For a discussion of the labour bureaux in the early 1980s, see Stanley Greenberg and Hermann Giliomee, 'The underbelly of privilege', in Hermann Giliomee and Lawrence Schlemmer, (eds.), *Up Against the Fences : Passes, Poverty and Privilege.* Cape Town: David Phillip, 1985, pp. 68 – 84.

12. For a discussion see Deborah Posel, 'Coloured Labour Preference Policy during the 1950s in the context of the National Party policy on African urbanisation', paper presented to a conference on the Western Cape, University of Cape Town, 1986; Gavin Lewis, *Between the Wire and the Wall, A History of South African 'Coloured politics'.* Cape Town: David Phillip, 1987, pp. 171, 219 – 20.

13. R P 38/1976 Report of the Commission of Inquiry into matters relating to the Coloured population group (Theron report), pp. 90 – 1.

14. Philip Frankel, 'The politics of passes : control and change in South Africa', *Journal of Modern African Studies*, 17, 1979, p. 206.

15. Simon Bekker and Richard Humphreys, 'Continuity and change in Administration Board regulations', paper presented to the conference on Economic Development and Racial Domination, University of Western Cape, 1984, pp. 16 – 19.

16. Charles Simkins, *Four Studies on the Past, Present and Possible Future of the Distribution of the Black Population in South Africa.* Cape Town: Saldru, 1983.

17. Francis Wilson, 'Mineral wealth and rural poverty : an analysis of the economic foundations of the political boundaries in South Africa', in *Up Against the Fences*, p. 56.

18. S. F. Frankel, *Investment and the Return to Equity Capital in the South African Goldmining Industry.* Oxford: Blackwell, 1967, pp. 7 – 9, 45, cited by Lipton in *Capitalism and Apartheid*, p. 115.

19. Francis Wilson, *Labour in South African Gold Mines.* Cambridge: Cambridge University Press, 1972, pp. 66, 76 – 7. Lipton, *Capitalism and Apartheid*, pp. 115 – 6; Merle Lipton, 'Men of two worlds', *Optima*, 29, 3, 1980, p. 108.

20. Cited by Oliver Walker in *Kaffirs are Lively.* London: Victor Gollancz, 1948, p. 22.

21. Cited by Stanley Greenberg in *Race and State in Capitalist Development.* New Haven: Yale University Press, 1980, p. 167.

22. Cited by Lipton, *Capitalism and Apartheid*, p. 128.

23. T. Gregory, *Ernest Oppenheimer and the Economic Development of Southern Africa.* Cape Town : Oxford University Press, 1962, pp. 573 – 81.

24. Wilson, *Labour in the South African Gold Mines*, p. 55.

25. Cited by Colin Bundy, 'The abolition of the Masters and Servants Act', in A Paul Hare, et al., (eds.), *South Africa : Sociological Analyses.* Cape Town: Oxford University Press, 1979, p. 378.

26. Lipton, *Capitalism and Apartheid*, p. 118.

27. Jill Nattrass, *The South African Economy.* Cape Town: Oxford University Press, 1981, p. 120.

28. Lipton, *Capitalism and Apartheid*, pp. 104 – 5.

29. Greenberg, *Race and State in Capitalist Development*, pp. 95 – 6.

30. See Cosmos Desmond, *The Discarded People.* Baltimore : Penguin Books, 1971.

31. M. J. de Klerk, 'The Labour Process in Agriculture : Changes in maize farming during the 1970s', *Social Dynamics*, vol. 11, no. 1, 1985, pp. 7 – 31.

32. Lipton, *Capitalism and Apartheid*, p. 90.
33. Surplus Peoples Project, *Forced Removals in South Africa*. Johannesburg: Ravan Press, 1985. p. 75.
34. Lipton, *Capitalism and Apartheid*, pp. 96 – 7.
35. *The Citizen*, 26 September 1984.
36. Greenberg, *Race and State*, p. 181.
37. Greenberg, *Race and State in Capitalist Development*, pp. 191 – 208; R. Davies, 'Capital restructuring and the modification of the racial division of labour in South Africa', *Journal of Southern Africa Studies*, vol. 5, no. 2, 1979.
38. Greenberg, *Race and State*, pp. 181 – 2.
39. Colin Bundy, 'The abolition of the Masters and Servants Act', in A. Paul Hare, et al. (eds.), *South Africa: Sociological Analyses*. Cape Town: Oxford University Press, 1979, pp. 373 – 9.
40. Greenberg, *Race and State*, p. 160; Lipton, *Capitalism and Apartheid*, pp. 165 – 76.
41. Pam Christie and Colin Collins, 'Bantu education: apartheid ideology and labour reproduction', in Peter Kallaway, (ed.), *Apartheid and Education: The Education of Black South Africans*. Johannesburg: Ravan, 1984, p. 180. For a detailed discussion see Muriel Horrell, *Bantu School Education to 1968*. Johannesburg : South African Institute of Race Relation, 1968.
42. *HAD*, 1949, cols. 306 – 11.
43. *HAD*, 1950, col. 2534.
44. Gwendolen Carter, *The Politics of Inequality: South Africa Since 1948*. London: Thames and Hudson, 1958, p. 76.
45. Roger Omond, *The Apartheid Handbook*. Harmondsworth: Penguin, 1985, pp. 28 – 9.
46. J. J. F. C. Heydenrych, 'Die maatskaplike implikasies by die toepassing van Artikel 16 van Wet 23 van 1957', MA dissertation, University of Stellenbosch, 1968, p. 142.
47. Dian Joubert, *Met Iemand van 'n Ander Kleur: Beskouing en Wetgewing oor Ontug*. Cape Town: Tafelberg, 1974, pp. 90 – 2.
48. *Die Burger*, 21 February 1950.
49. John Dugard, *Human Rights and the South African Legal Order*. Princeton: Princeton University Press, 1978, pp. 61 – 2.
50. Robin Cohen, *Endgame in South Africa: The Changing Structures and Ideology of Apartheid*. London: James Currey,1986, p. 35.
51. Jack Simons in his preface for G. Watson, *Passing for White: A Study of Race Assimilation in a South African School*. London: Tavistock Publications, 1970, p. viii.
52. *HAD*, 1950, cols. 2519 – 21.
53. *HAD*, 1956, cols. 5259 – 60.
54. *SA Institute of Race Relations Survey*, 1956, pp. 36 – 8.
55. Cited by M. J. Mittner, 'Die burger en die Kleurling-politiek, 1948 – 1961', MA dissertation, University of Cape Town, 1986, p. 121.
56. *HAD*, 1950, col. 2530.
57. Cited in Dugard, *Human Rights and the SA Legal Order*, p. 62.
58. Stanley Uys, 'Racial love at the end of the line', *Guardian Weekly*, 7 September 1974, reprinted in John Western, *Outcast Cape Town*, Cape Town. Human and Rousseau, 1981, pp. vi – vii.
59. *HAD*, 1950, A. J. R. van Rhyn, col. 9686.
60. *HAD*, 1950, cols. 7722, 7726.
61. For detailed analyses see K. Kirkwood, *The Group Areas Act*. Johannesburg: South African Institute of Race Relations, 1955; M. Horrell, *The Group Areas Act... its effects on human beings*. Johannesburg: S A Institute of Race

Relations, 1956. For a superb analysis of the impact of Group Areas on a particular city, see Western, *Outcast Cape Town*.

62. Carter, *The Politics of Inequality*, p. 98.
63. Carter, *The Politics of Inequality*, p. 96
64. *HAD*, 1950, col. 9453.
65. *HAD*, 1950, cols. 7724 – 5.
66. *HAD*, 1950, col. 7745.
67. For a full discussion see Western, *Outcast Cape Town*, pp. 103 – 34.
68. *Sunday Tribune*, 2 October 1988, p. 4.
69. Paul Bell, 'Our only Sin', *Leadership*, vol. 7, no. 3, 1988, p. 57.
70. The paragraphs on Indian removals are based on Anthony Lemon, *Apartheid: A Geography of Separation*, Westmead: Saxon House, 1976, pp. 105 – 22.
71. T. R. H. Davenport, *South Africa : A Modern History*. Johannesburg: Mac-Millan, 1987, p. 548.
72. Lemon, *Apartheid: A Geography of Separation*. pp. 75 – 6.
73. Gillian Hart, *African Entrepreneurship*, Occasional Paper, no. 16, Institute of Social and Economic Research, Rhodes University, Grahamstown; Leo Kuper, *An African Bourgeoisie: Race, Class and Politics in South Africa*. New Haven: Yale University Press, 1965, ch. 17.
74. *HAD*, 1959, cols. 3259, 3263 – 4.
75. *HAD*, 1959, col. 8318.
76. H. F. Dickie-Clark, 'The dilemma of education in plural societies : the South African case', in A. Paul Hare, et al., (eds.), *South Africa: Sociological Analysis.* Cape Town: Oxford University Press, 1979, p. 174.
77. For an elaboration of this theme see the chapters by Lawrence Schlemmer and F. van Zyl Slabbert in Peter Randall, (ed.), *Towards Social Change: Report of the Spro-cas Social Commission.* Johannesburg: Christian Institute, 1971, pp. 5 – 71.
78. For a detailed discussion of the Broederbond see Ivor Wilkins and Hans Strydom, *The Super-Afrikaners : Inside the Afrikaner-Broederbond.* Johannesburg : Jonathan Ball, 1978; J. H. P. Serfontein, *Brotherhood of Power : An Exposé of the Secret Broederbond.* London: Rex Collings, 1979.
79. For a full discussion see Leonard Thompson, *The Political Mythology of Apartheid.* New Haven: Yale University, 1985.
80. J. M. du Preez, *Africana Afrikaner : Master Symbols in South African School Textbooks.* Alberton: Librarius, 1983.
81. Heribert Adam, (ed.), 'The South African power elite', in Heribert Adam, (ed.), *South Africa : Sociological Perspectives.* London: Oxford University Press, 1971, p. 80.
82. Hennie Kotze, 'Ja-broers in niemandsland', *Die Suid-Afrikaan*, Summer 1987, pp. 29 – 31.
83. *Apartheid and Guardianship : Short summary of NP policy.* Published by HNP Head Office, Cape Town, c. 1948.
84. For a good exposition of the thinking of the Cape wing of the NP, see Mittner, 'Die burger en die Kleurling-politiek, 1948 – 1961'.
85. Henry Kenney, *Architect of Apartheid : H. F. Verwoerd – An Appraisal.* Johannesburg: Jonathan Ball, 1980, p. 78.
86. *HAD*, 1959, col. 6237.
87. 'Coloured viewpoint', column first published in *Cape Times*, 10 March 1960, reprinted in R. E. van der Ross, *Coloured Viewpoint.* Belville: University of Western Cape, 1984, p. 102.
88. 'Coloured viewpoint', 24 August 1961, reprinted in Van der Ross, *Coloured Viewpoint,* p. 181.

89. Van der Ross, *Coloured Viewpoint*, p. 181.
90. *HAD* (Afrikaans edition), 1965, cols. 4403 – 4410.
91. J. J. J. Scholtz, (compiler), *Vegter en Hervormer : Grepe uit die Toesprake van P. W. Botha.* Cape Town: Tafelberg, 1988, pp. 52 – 3.
92. *Survey of Race Relations in South Africa, 1978.* Johannesburg: SA Institute of Race Relations, 1979, p. 321.
93. W. H. Thomas, 'The Coloured people and the limits of separation', in Robert Schrire (ed.), *South Africa : Public Policy and Perspectives.* Cape Town, Juta, 1982, p. 159 – 60.
94. Cited by Edgar Brookes, *Apartheid : A Documentary Study of Modern South Africa.* London: Routledge and Kegan Paul, 1968, p. 149.
95. J. J. J. Scholtz, 'Apartheid', *Die Burger*, 19 November 1985.
96. Thomas, 'The Coloured people and the limits of separation', p. 160.
97. Brian Urquhart, *Hammerskjöld.* London: Bodley Head, 1973, p. 498.
98. C. E. W. Simkins, 'Agricultural production in the African reserves of South Africa, 1918 – 1969', *Journal of Southern African Studies*, 1981, pp. 256 – 83.
99. Servaas van den Berg, 'An overview of development in the homelands', *Up Against the Fences*, p. 198. For an important study of the economics of homelands development, see Italo Trevisan, 'Independent Homelands : an analysis of selected issues in South Africa-Homeland relations', MA dissertation, University of Cape Town, 1984.
100. This is discussed in detail in Heribert Adam and Hermann Giliomee, *Ethnic Power Mobilized : Can South Africa Change?* New Haven: Yale University Press, 1979, ch. 6.
101. Norman Bromberger, 'Government policies affecting the distribution of income 1940 – 1980', in Schrire, *South Africa : Public Policy and Perspectives.* pp. 172 – 4.
102. Bromberger, 'Government policies', p. 175.
103. Bromberger, 'Government policies', p. 177 – 86.
104. Michael McGrath, 'The racial distribution of taxation and government expenditures', University of Natal paper, 1979.
105. McGrath, 'The racial distribution'. McGrath assumed that maximum redistribution to black groups had taken place.
106. *HAD*, 1948, col. 200.
107. Anthony Lemon, *Apartheid in Transition.* Aldershot: Gower, 1987, p. 205.
108. John Rex, 'The plural society : the South African case', *Race*, XII, 4, 1971, p. 405.
109. R. J. Davies, 'The spatial formation of the South African city'. *Geojournal*, supplementary issue 2, 1981, p. 59 – 72.
110. This is the theme of John Vaizey, *Scenes from the Institutional Life and other Essays.* London: Weidenfeld and Nicholson, 1982.
111. Breyten Breytenbach, 'The alienation of White South Africa', Alex la Guma, (ed.), *Apartheid.* New York: New International Publishers, 1971, p. 142.

CHAPTER 4

Reform-apartheid

Although apartheid did have an extraordinary appeal for Afrikaner nationalists in the 1950s and 1960s, it has always been the instrument of Afrikaner nationalism, never its master. It was, and still is, designed to serve the interests of the dominant classes within Afrikanerdom and the white community at large. These interests started to change in the 1960s. In constructing and later moving away from apartheid, the Afrikaner nationalists are no different from any other ruling group. In a comparative study the American sociologist, Pierre van den Berghe, observed that ruling groups, whether ethnically defined or not, act 'in the furtherance of their interests, or at least in what they perceive to be their interests'. According to Van den Berghe, ruling groups differ not so much 'in their degree of benevolence, but in the extent to which they misjudge their interests'.[1]

The ruling group in South Africa acted fundamentally according to its perceived interests in embarking on its reform initiatives. However, these initiatives cannot, by any stretch of the imagination, be described as the outcome of a coherent or consistent programme. One should rather speak of a change process comprised of various elements, some of them mutually contradictory. Three main elements or characteristics can be distinguished. Firstly, there have been *reform initiatives* aimed at curbing explicit white exclusivity and privilege, and at the substitution of technocratic and free-market principles for racially based criteria. Secondly, there has been an *elaboration or refinement of apartheid* which has seen concepts such as 'own affairs' being used to give substance to the apartheid idea that the four statutory communities should administer themselves. Thirdly, there has been an *unravelling of the system* in areas such as influx control. Inexorable pressures forced the apartheid state to abandon policy parameters and fall back on a set of *ad hoc* arrangements. This has resulted in ideological confusion and a 'bureaucratic tangle' which has seen top bureaucrats and lower-level officials at odds with one another over the content and thrust of government policy.[2] In this chapter we shall briefly touch on all these themes.

Forces propelling the change process

The elements which have propelled a shift away from classic apartheid include the white demographic decline, growing black militancy, foreign pressure, changes in the Afrikaner class composition, and the fiscal crisis of the South African state. Perhaps the most important of these is the changing demographic profile of South Africa.

During the first half of the twentieth century South African whites had, in terms of their relative population share, enjoyed strategic self-sufficiency. This meant that it had not been necessary for them, as it had been for the Brazilian colonial state in the nineteenth century, to free large numbers of blacks in order to fill the intermediate jobs in the economy and administrative apparatus.

By 1960 an important change had begun to make itself felt. In the preceding fifty years whites constituted 20 per cent of the total population (Table 4.1), but after 1960 the white demographic base started to shrink. The proportion of whites to the total population fell to 15 per cent by 1985 and is projected to decline to 10 per cent by the year 2005. As a result, an acute shortage of white manpower has manifested itself in both the public and the private sectors.

TABLE 4.1: TOTAL POPULATION BY STATUTORY GROUPS, 1951 – 87

	1951	1970	1980	1987
African	8 560 083	15 057 952	21 307 749	*26 313 898
White	2 641 689	3 752 528	4 453 273	*4 911 000
Coloured	1 103 016	2 018 453	2 554 039	*3 069 000
Asian	366 664	620 436	794 639	*913 000
Total	12 671 452	21 448 169	29 109 700	*35 206 898

*Source: *Race Relations Survey.*

In the civil service during the 1950s and 1960s, the apartheid state greatly increased the number of whites employed to staff the apartheid bureaucracies and implement the myriad of apartheid laws. By the mid-1970s these bureaucracies had reached their logistical limits. The state had over-reached itself in terms of its own spending capacity and there were simply not enough whites to staff the upper levels of both the private and public sectors. The mammoth Department of Bantu Administration and Development found itself incapable of stemming the flow of Africans to the cities. The large numbers of urbanizing Africans, together with the rising cost of building materials, led to the growth of huge squatter-camps, mostly within the homeland borders. By the late 1970s, half a million people were housed in informal settlements in greater Inanda adjoining Durban, 250 000 in Edenvale-Zwartkops near Pietermaritzburg, 300 000 in Winterveld near Pretoria, 100 000 in Mdantsane near East London, and approximately the same number in Crossroads near Cape Town's airport.

Increasingly the government looked to the various black groups to take over the administration of their own communities (and in the case of Africans, to implement influx control). It also wanted blacks to supplement the over-extended police and defence forces. As President Botha expressed it somewhat crudely in a biography:

> I realise that there are today tens of thousands and hundreds of thousands of brown people who are in all respects better than the weakest [*swakste*] whites. This is one of the burning questions of our population. We must give these people a say [*inspraak*], a greater share in the land — in the administration of the country, as civil servants, and in many other positions which you cannot fill with your weak whites... If we don't do this we will cause our own down-fall.[3]

Certain developments facilitated the entry of blacks into strategic levels of the economy. Firstly, the economy has moved further away from the early phase of industrialization, characterized by the predominance of mining and agriculture which relied heavily on uneducated and poorly-trained black workers, many of whom were migrants. Since the Second World War the rise of the manufacturing sector (Table 4.2) has meant an increasing demand for a skilled and productive work-force, settled on a family basis in the cities.

The relative demographic decline of whites has forced industrialists to turn to blacks for skilled labour. After the rapid economic growth of the 1960s, whites were no longer able to supply fully the demand for skilled labour. Between 1971 and 1977, whites comprised only a quarter of the increase in fully employed, skilled, blue-collar workers

(15 600 out of 65 700). Increasingly, employers wanted to train blacks and employ them in more senior positions. In many cases they were blocked by white trade unionists who attempted to reassert the privileged position of the white worker. This gave rise to strains in the production process. In 1977, at the end of a severe recession, 45 per cent of a sample of leaders in the manufacturing sector expressed the view that difficulties in acquiring adequate skilled labour were causing bottle-necks in production; by 1980 this figure had risen to 80 per cent. These shortages have forced both the government and employers to open the labour market.

TABLE 4.2: CONTRIBUTION (%) OF SECTORS TO THE NATIONAL ECONOMY (GDP) AT CONSTANT 1974 PRICES BY TYPE OF ECONOMIC ACTIVITY

	Agri-culture & forestry	Mining and quarrying	Manu-fac-turing etc.	Commerce, catering	General govern-ment	Other
1950	14,5	15,1	16,5	11,3	11,2	31,4
1960	11,9	17,8	19,1	10,7	10,8	29,5
1970	8,3	18,3	21,2	12,2	9,5	30,5
1980	8,5	11,5	25,6	12,1	9,9	32,4

Sources: 1. Stanley Greenberg, *Race and State in Capitalist Development*. New Haven: Yale University Press, 1980, p. 426.
2. *South African Reserve Bank Quarterly Bulletin*, 1981.

The result of these developments has been a major change in the racial profile of the labour market. By the beginning of the 1960s whites still made up 82 per cent of middle-level manpower. This had dropped to 65 per cent in 1981. The entry into middle-level positions has been spear-headed by coloured and Indian workers, but the advance of Africans has not been insignificant (Table 4.3). In the past, blacks entered jobs vacated by whites who had moved upwards. This left the apartheid hierarchy intact. However, the changes in the labour market over the past few decades have gone beyond this ratchet-like pattern. Merle Lipton wrote as early as 1974:

> Increasing numbers of them [blacks] in many branches of industry... are doing the same jobs as whites, even though this may be done on different shop floors of the same factory, in different factories or in different areas. More-over, many of these blacks.... are in superior jobs to those of some whites.[4]

TABLE 4.3: AFRICAN OCCUPATIONAL ADVANCEMENT

Sector	Africans as a proportion of all employees in the higher-level occupations	
	1970	1980
Professional, higher-level technical, and related workers	21%	29%
Administrative, managerial and clerical workers	18%	27%

Source: Centre for Applied Social Sciences, University of Natal, Durban.

In order to meet the growing demand for skilled labour, the government has greatly expanded black, and particularly African, education over the past two decades. The results can be seen in Table 4.4.

TABLE 4.4: AFRICAN ENROLMENT IN HIGHER EDUCATION SINCE 1960

Year	Secondary school	Highest school standard	University students
1960	54 598	717	1 871
1965	66 568	1 606	1 880
1970	122 489	2 938	4 578
1975	318 568	9 009	7 845
1980	577 584	31 071	10 564
1985	*1 192 932	*34 733	*49 164

Sources: 1. Colin Bundy, 'Schools and Revolution', *New Society*,
       10 January 1986.
    2. **Race Relations Survey*, 1986.

When the system of apartheid was introduced in the 1950s, the majority of the African population was illiterate or barely literate, and performed unskilled work. As a result, it was relatively easy for the apartheid state to deny such a population their political rights and deflect their aspirations to distant 'homelands'. By comparison the African population of the mid-1980s is far better educated and trained, and much more militant. Surveys have consistently shown that the higher the level of education of African children, the more acute is their political discontent and the more pressing are their political and status demands.

The results from a survey of 300 Africans in Natal and KwaZulu in 1981 are particularly telling (Table 4.5). These results relate to acceptance of the segregated order in South Africa. A number of issues were presented to respondents who were then asked, 'Which of the following things would you be happy for whites to keep separate for themselves in South Africa?'

TABLE 4.5: AFRICAN POLITICAL RESPONSES (%) ACCORDING TO EDUCATION

	Level of education			
Whites can have their own:	Std 2 or below	Std 3 – 6	Std 7 – 9	Std 10 or above
Laws against mixed marriage	70	65	45	18
Housing areas	62	52	32	15
Schools	53	34	26	13
Farmlands	47	38	29	11
Recreation facilities	41	27	13	9
Transport and buses	36	26	18	2

Note: Only the percentages accepting separation are given.

From the table it is evident that there is a dramatic divergence of views according to the level of education of the respondents. For instance, the poorly-educated migrant workers desire land particularly strongly (note for example the result on farmlands) but they are not that concerned about status demands. In contrast, the African educated élite rejects apartheid practices which deny them their claim to full equality.

Rising black militancy has been another major spur to the change process. It was on the shop floor that blacks first flexed their muscles. In 1973 the industrial peace which South Africa had enjoyed as a result of severe repression was shattered by the Durban strikes. During the 1960s, the number of black workers involved in industrial disputes had never been higher than 10 000 per year. This figure suddenly jumped to 100 000 in 1973. After the Durban strikes, Prime Minister Vorster declared that the huge racial wage gap had to be closed and admitted that 'these events contained a lesson': black workers had to be treated not as 'labour units, but as human beings with souls'.[5] As late as 1970 the government had pledged to maintain the 'civilized labour policy' which reserved the skilled, better paid jobs for whites; now, however, the government urged employers to narrow the wage gap and undertook to set an example in the civil service.

Initially both the government and employers had tried to repress the independent trade union movement which took root in the 1970s. However, the numerous wildcat strikes and the growing authority vacuum on the shop floor forced both the government and employers to seek some accommodation to restore a measure of stability. The political uprisings of 1976 and 1984 – 6, which were accompanied by several mass stay-aways, contributed to the complex set of pressures which forced the government to abandon some of the other manifestations of apartheid.

The dynamics of economic growth and class diversification also impacted on the Afrikaner ruling group and its perceptions of apartheid. When the apartheid policy was introduced in the 1950s, the majority of Afrikaners were insecure blue-collar workers and marginal or struggling farmers, dependent on the state to maintain their living standards and social privileges. By the mid-1970s, however, at least 70 per cent of Afrikaners belonged to a relatively secure middle class (Table 4.6). This middle class began to find it increasingly difficult to reconcile the blatant racism and discrimination of classic apartheid with their professional norms and values. Instead of abandoning the National Party, this middle class insisted, at the very least, on a less blatant form of apartheid.

TABLE 4.6:   % OF AFRIKANERS IN BROAD CATEGORIES OF OCCUPATION, 1936 – 77

Category	1936	1946	1960	1970	1977
Agricultural occupations	41,2	30,3	16,0	9,7	8,1
'Blue collar' and other manual occupations	31,3	40,7	40,5	32,4	26,7
'White collar'	27,5	29,0	43,5	57,9	65,2
	100,0	100,0	100,0	100,0	100,0

Source : J. L. Sadie in H. Giliomee and H. Adam, *Afrikanermag : Opkoms en Toekoms*. Stellenbosch, UUB, 1981, p.130.

Another major force for change has been the pressure from the outside world, which looks upon apartheid as an obnoxious reminder of the West's colonial past and a totally unacceptable violation of basic human rights. Foreign pressure has come from many sides and has assumed many different forms. The most sustained and probably one of the most effective sanctions campaigns has been Africa's near exclusion of South Africa from the markets lying to its north. This has dealt a major blow to the manufacturing sector which, by the early 1980s, contributed only 10 per cent to South Africa's export earnings. In order to open up this market both B. J. Vorster and P. W. Botha attempted to start a dialogue with African states. Both removed some of the cruder aspects of apartheid in order to facilitate this process, but neither met with too much success. The trade with Africa has grown annually at a rate of 7 per cent in recent years but there has not been a major breakthrough in penetrating markets north of the border. Another long-standing sanction relates to international sport. South Africa's efforts to ward off sport isolation have also contributed to minor policy shifts.

The removal of the protective belt of white-ruled states on the northern and eastern borders between 1974 and 1979, as well as the perceived Communist-inspired onslaught, gave rise to a growing realization that naked white oppression had to be replaced by a system that would attract black allies. As a senior member of the Afrikaner press formulated it: 'We cannot fight on the borders and continue to play the constable over blacks in our country'.[6]

Another pressure on government is the economy, which is wide open to the world economy with exports and imports amounting to more than 60 per cent of GDP. As a developing country, South Africa needs to import capital. It has been estimated that the country has to attract between 7 and 10 per cent of its capital needs from abroad in order to maintain a GDP annual growth rate of 5 per cent. If direct, private, foreign investment had been terminated, the growth rate of the economy between 1974 and 1983 would have been lowered by approximately 24 per cent.[7]

Since 1976 continued disinvestment has been dependent upon whether South Africa was prepared to meet two requirements. On the one hand the Western world demanded a more humane system than apartheid and, on the other, the multinationals demanded greater stability. Foreign investors were scared off by the worsening security situation and rising black wages. American multinationals have also faced growing pressure from universities, churches, and city and state authorities to disinvest from South Africa. By the end of the 1970s, new direct investment by American multinationals had virtually ceased.

These pressures finally resulted in the multinationals in South Africa, as well as many South African companies adopting the Sullivan and other employers' codes. In these codes they pledged themselves to improve the working conditions of black employees, abolish discrimination in the work place, assist black empowerment and promote the transformation of the South African social structure. The government also became sensitive to these pressures. Sanctions, boycotts, and disinvestment played a significant role in the decision of government to liberalize the labour laws.

By the mid-1980s, loans from foreign banks had come to form an acute pressure point. A considerable increase in the debt-to-GDP ratio occurred owing to the fact that the economy had slowed down from an average of 5 per cent in 1964 – 74 to below 3 per cent in 1974 – 84, and to only 1,2 per cent in 1980 – 84. Whereas foreign debt had represented only about 20,5 per cent of GDP in 1980, this figure jumped to 38,2 per cent in 1984.[8] A relatively high proportion (60 per cent) of this foreign debt comprised short-term loans. This vulnerability forced the South African government to consider bankers' demands for further reforms sympathetically, rather than responding with an iron fist to the uprising which broke out in September 1984. However, in August 1985 a number of foreign banks called in their loans and this contributed to the government's decision to resort to full-scale repression of the uprising. This decision was later modified and the government promised further reforms in order to salvage trade credits and woo back foreign bankers.

Finally, the government is being pressurized by its continuing fiscal crisis. This has its roots in a stagnant economy, a burgeoning public sector (state spending as a proportion of GDP rose from below 30 per cent in the late 1970s to over a third in the mid-1980s), and a narrow tax base. The fiscal crisis has fuelled the growth of the white right, since the individual white taxpayer is increasingly footing the state's bill. In 1981, company tax represented 41 per cent of the total tax burden, but by 1988 this had dropped to just over 20 per cent. By contrast, personal tax rose from 17 per cent of state revenue, to 31,4 per cent in the same period and it currently represents the largest source of state revenue, followed by sales tax at 26,9 per cent. In 1988, whites paid more than 90 per cent of the total personal tax collected. Studies have shown that a person in the middle income category, earning R26 000 a year in 1986, paid three times the level of personal tax such an income would have attracted in 1981.[9]

These trends have been grist to the mill of the Conservative Party (CP). It campaigns effectively on the platform of a white populism that attacks both government spending on blacks and the avarice of big business. *Die Burger* reported on 13 October 1988 that Dr A. P. Treurnicht, leader of the CP, told audiences that blacks 'which command 45 per cent of the spending power paid only 8,6 per cent of the income tax'. He clearly implied that this was an intolerable situation. These attacks have hit the government at its most sensitive spot, namely electoral support. As a result it has tried to curb state spending and to deregulate the economy. It has also sought to extract more revenue from blacks.

Reform, adaptations and elaborations

We shall firstly discuss changes in communal apartheid and the regulation of labour before moving on to an analysis of the adaptations to political apartheid. The first steps away from communal apartheid were prompted by South Africa's increasing international isolation and began in the fields of international sport and diplomatic relations. John Vorster, who succeeded Hendrik Verwoerd as Prime Minister in 1966, personally took the lead in this. As a minister in the Verwoerd cabinet, he had insisted that multi-racial sport would 'never' take place in South Africa and personally declined invitations to attend multi-racial receptions. However, apartheid in sport led to South Africa's exclusion from the Olympic Games after 1960, as well as the cancellation of the New Zealand rugby tour over the issue of Maori players in the visiting team.

In the NP's internal debates, Piet Koornhof argued strongly that if adaptations were not made to the sports policy, the (white) youth would 'suffocate'. Vorster, in private conversations with NP parliamentarians, suggested a new definition of apartheid. He stated that: 'The identity of each of the [statutory] groups must be preserved, friction points must be eliminated and each group must get the opportunity for development'.[10] The argument gained support on the grounds that, since no nation was inferior according to apartheid ideology, there could be no objection to mixed sport on an international level. The space that Vorster won through his reformulated definition of apartheid enabled him to push through a new sports policy which, after a tortuous fifteen-year process, finally introduced multi-racial sport and the scrapping of separate facilities for spectators.[11]

The establishment of diplomatic links with Malawi and the rise of a bureaucratic élite in the homelands set the precedent for allowing some blacks to live in white group areas and to use public facilities. The fiction of 'international' hotels and restaurants expedited the opening of these facilities to all blacks. The 1970s also saw the gradual disappearance of the most offensive features of apartheid, namely segregated entrances, lifts, waiting-rooms, toilets and benches in parks. Over the following fifteen years some beaches, parks, buses, trains, theatres and libraries were desegregated. Through the years the government has tolerated a few 'grey areas' (mixed residential areas) in Cape Town, Durban and Johannesburg, and in 1988 it formally approved the principle of free settlement areas. In terms of this, a suburb may be declared open after various requirements have been met, including the sounding out of the opinion of whites who may be affected. However, the state is increasingly prepared to turn a blind eye where there are no strong white objections to integration.

According to NP thinking, white identity and communal apartheid were not threatened by the elimination of these forms of segregation which concerned the public space in which the communities mingled. Once communal apartheid was firmly established, the government also permitted some desegregation in education. It allowed white universities and technikons to increase their intake of black students and subsequently dropped the idea of a quota system. Although the government still issues occasional warnings, universities are now free to decide on the admission of students. Since the beginning of the 1980s there has also been a more flexible policy with respect to the entry of blacks into private schools.

Despite these measures, the government has retained the cornerstones of communal apartheid, namely race classification, group areas and segregated education in state schools. In 1985 it abolished the racial sex laws (Article 23 of the Immorality Amendment Act and the Prohibition of Mixed Marriages Act), but it has ostracized whites who have married across the colour line by forcing them to live either in one of the black residential areas or in one of the open residential areas it has sanctioned. In 1989 the government, for the first time, raised the possibility of changes to the Population Registration Act. Senior cabinet ministers declared that the 'group character' of society remained the foundation of government policy. However, the NP was prepared to allow groups to form voluntarily and to entertain the possibility of an 'open group'.

A second major area of change has been in the regulation of labour. In the course of the 1970s the government began to remove restrictions on black labour and other black economic activities. As the white skilled-labour shortage worsened, the government became ever more impatient with white trade unions which were hampering the training of blacks and thus blocking black advances into skilled jobs. In 1973 it was announced that blacks, including Africans, could do skilled work in the white areas. The government did not rigorously adhere to its promise that it would consult with white trade unions before making this decision. In 1975 the defence force announced that black soldiers would enjoy the same status as whites of equal rank, and that whites would have to take orders from black officers. This broke the rule that the hierarchical structure (or ratchet) must be kept intact, with blacks always working under whites.[12] By the end of the 1970s, merit-based promotion was generally accepted in the private sector, but not commonly implemented. Nevertheless, a 1979 survey revealed that blacks supervised whites in one-fifth of a sample of companies. This trend is continuing.

Black education received a major boost in 1972 when the government abolished the Verwoerdian principle of pegging expenditure on African education to direct African taxes. Since that year state spending on African education has increased by more than 1 000 per cent. The racial gap in state spending has gradually shrunk, most dramatically in the per capita ratio of expenditure on white and African children, which narrowed from 16,6 : 1 in 1968, to 7,2 : 1 in 1983. In response to the De Lange panel report of 1983, the government committed itself to equalize, over time, state spending on education for the various statutory groups. However, after the imposition of economic sanctions in 1985 – 6, the government began to back down on this commitment.

The Wiehahn Commission report, published in 1979, set the stage for the government's most important labour reforms. In the following years the government accepted the proposals to scrap statutory job reservation, include black trade unions in the statutory industrial relations system, indenture black apprentices for training, advance the principle of equal payment for work of equal value and abolish segregation regulations under the laws relating to factories, shops and offices. Mobilized resistance by independent trade unions and powerful practical considerations have forced the government to move far beyond its original intention of creating carefully controlled, segregated African unions which excluded migrants. Mixed unions, with strong migrant labour representation, have made their presence felt on the new industrial relations scene.

By and large, the state has stuck to a policy of non-intervention in strikes. This means that it no longer considers itself responsible for restoring and maintaining industrial peace through policing action when the labour disputes are purely about wages and working conditions.[13] Trade union leaders advocating political action have, however, been subject to harassment and detention.

The easing of restrictions on black labour, coupled with the willingness of both the state and private employers to narrow the wage gap between the various race groups, has had a considerable impact on income distribution. Lipton gives a succinct account:

> From 1970 – 82, real wages for Africans in manufacturing and construction rose by over 60 per cent, compared with 18 per cent for whites; on gold mines real wages for

Africans quadrupled, while those for white miners rose by 3 per cent; on white farms real wages for Africans doubled between 1968 – 69 and 1976. As a result, the white share of personal income declined from under 70 to under 60 per cent, while the African share rose form 19 to 29 per cent.[14]

Yet by the end of the 1970s African ownership of assets was still extremely limited. They were still subjected to the humiliating pass laws, and they enjoyed virtually no political representation on any level of government outside of the homelands. In the case of coloured people and Indians, political representation was extremely limited and shortages of urban housing had alienated many. By the mid-1970s, and especially after the Soweto uprising of 1976, enlightened businessmen and government spokesmen were increasingly stressing the need to give blacks 'a stake in the system'. They argued that the system must be opened 'to save it and build on it'.[15] The race was on to develop a black middle class that could stabilize the system.

In line with this objective the government eased the restrictions on African businessmen in townships and also allowed them to trade in certain central business districts. By the end of 1986 integrated business districts had been established in fifty-five municipalities. The government also finally allowed Africans to acquire immovable property in townships. The housing policy underwent many contortions. At first Africans were allowed to acquire houses on a thirty year, and then on a ninety-nine year leasehold, but finally, it was decided in 1986 to grant them freehold home ownership. To encourage African home ownership, the government put 400 000 houses up for sale at bargain prices.

It has also attempted to ease the restrictions on the movement of Africans. After numerous botched attempts, the pass laws were finally abolished in 1986. However resident citizens of the independent homelands are subject to 'passport control' since they are considered aliens who are not allowed to take up employment without permission. Anyone who provides them with employment or accommodation is subject to heavy penalties. In practice the bureaucratic inertia which has set in since the early 1980s has allowed these people to seek work without too much harassment. The government has begun to make land more freely available and one expects that over the medium term the covert forms of influx control will be swept aside by the sheer number of migrants, as was the case with the pass laws and the other more overt influx controls.

The past two decades have also seen the introduction of a set of political and administrative adaptations. One fundamental reform has been the repeal in 1985 of the prohibition of mixed political parties. Surprisingly, this has had little practical effect. Most of the other initiatives cannot be called the reform of apartheid but rather the elaboration of apartheid. Three key apartheid principles have remained inviolate. Firstly, there is unwavering determination to protect Afrikaner, and larger white, material interests and identity needs. Security legislation has introduced even more draconian controls than those passed in the early 1960s. Secondly, the government has continued to organize political life in terms of relations between the four statutory apartheid groups. The government has clung to the view that group political 'autonomy' and development are prerequisites for good government. Thirdly, the government has revived the idea of a 'constellation' or 'commonwealth' of states, first proposed by H. F. Verwoerd. This promotes close co-operation and links between the apartheid groups, regions or states. We shall briefly discuss the elaboration of the homelands policy and the constitutional initiatives in the common area.

In 1972 Transkei was the only self-governing state but the other homelands were advancing rapidly down the same path. By January 1977 Bophuthatswana, Ciskei, Lebowa, Venda, Gazankulu, QwaQwa, and KwaZulu had all become self-governing. In 1976 Transkei became independent, followed by Bophuthatswana (1977), Venda (1979) and Ciskei (1981). One of the main aims of the South African government in granting independence was to deprive as many blacks as possible of their citizenship of South Africa. In 1978 Dr Connie Mulder gave the classic exposition of this policy goal:

> If our policy is taken to its logical conclusion as far as black people are concerned, there will be not one black [meaning African] man with South African citizenship... Every black man in South Africa will eventually be accommodated in some independent new state in this honourable way and there will no longer be a moral obligation on this Parliament to accommodate these people politically.[16]

In 1979 Botha announced the concept of a constellation of states within the southern African region. The idea was generally hailed as a new departure in government policy, although it had been in the air for the past two decades. Verwoerd envisaged that once the homelands had become independent, the relationships between them and South Africa proper would be comparable to those existing between members of the British Commonwealth. Vorster spoke of a 'power bloc' or constellation of states, maintaining close economic ties with one another. What was new in the approach taken by P. W. Botha was the implicit suggestion that the homelands could become the building-blocks of a federation which would 'politically reunite' South Africa. The Verwoerdian goal of separate, viable homeland economies was abandoned. In its place, Botha announced a new regional economic strategy which would involve 'economic development co-operation transcending the borders of the Republic and the homelands'. Balancing growth-points would be developed in the regions to counteract the powerful attraction of the large metropolitan areas.

Reversing the policy of steadily denationalizing all Africans, as outlined above by Mulder, the government in 1985 and 1986 finally acknowledged the permanence of 'non-independent homeland' Africans in the white areas and accepted their claim to full citizenship. In a booklet published in 1985 entitled '... *And what about the black people?*', the government acknowledged that the political linkage of Africans to the homelands was no longer feasible for the majority of Africans, and that negotiations would have to take place with their leaders about the form in which their political rights might be realized.

These initiatives largely consigned the constellation idea to the back burner, if not to oblivion. Nevertheless, the guide-lines for negotiations, as spelt out in this booklet, amounted to a modernized form of apartheid. They included the following government demands: the 'group nature' of South African society would have to serve as the point of departure for any future political dispensation; no group should be dominated by another group; the self-determination of each group over its own affairs and matters of common concern was central; and each group should have its own community life with its own residential areas and its own schools.[17]

As a result of reform-apartheid, South Africa presently has a political system which, in the multiciplity of its ethnic structures, is unique. A University of Cape Town sociologist, Michael Savage, has summarized the bureaucratic ramifications of reform-apartheid:

> The South African political system has given birth to 13 Houses of Parliament or Legislative Assemblies, as well as the President's Council with quasi-legislative functions. There are three legislative chambers in the Central Parliament, six Legislative Assemblies in what are termed the 'non-independent black states', and four legislative Assemblies in the 'independent states'.

beer-halls had been a major source of revenue for the Administration Boards which had administered the townships. However, many of these halls were destroyed by the youth in the uprising of 1976 and subsequently the government had begun phasing out this source of revenue. The raising of rents and levies by the councils, which were themselves considered to be illegitimate by large numbers of Africans, sparked off an uprising in the Vaal triangle in September 1984. This soon escalated into a major black rebellion across the country, which lasted until the end of 1986. Although numerous town councillors were forced to resign by their communities, the government estimated that, by the end of 1987, more than 80 per cent of the elected township leaders were once again in office.

By the mid-1980s the government had abandoned its efforts to treat African local government as a distinct constitutional category. It was now presented as an integral part of the local government machinery of the Republic. The Administration Boards were abolished and African local government was transferred to the provinces. The all-white provincial councils were also abolished and control over the provinces was placed in the hands of government-appointed executives, which included Africans. Finally, Africans, as well as other groups, are represented on Regional Services Councils (RSCs) which are used to transfer financial resources for infra-structural development in townships (see pp. 142 – 5).

The Tricameral Parliament, which we shall discuss in more detail later, is based on the system of 'own affairs'. It should be emphasized, however, that the 'own affairs' system is the culmination of the apartheid concept rather than its reform. There is no question of segmental autonomy as prescribed by consociation theorists. The coloured and Indian Houses of Parliament do not generate their own revenue, and there is no proportional allocation of funds. The financial strings are still firmly held by the President and his cabinet.

The administration of central government (as distinct from own affairs administrations) is still nearly all-white. Out of a total of 10 966 state employees in the eight upper levels of the general government sector, 96 per cent are white, 1,8 per cent African, 1,4 per cent Indian, and 0,7 per cent coloured. Only six out of twenty-six departments employ a black person in the five most senior posts. In none of the departments are blacks represented in the top three levels. Since 1984 the government has appointed two black ambassadors. The most senior position held by a black in the defence force is that of lieutenant-colonel. The first black judge was appointed to the bar in 1987.[20]

Occupying seats in these 14 bodies are 1 270 members consisting of 308 Members of the 3 Houses of the Central Parliament; 60 members of the President's Council; 501 members of the Legislative Assemblies of the 'non-independent black states' and 401 members of the 'independent black states' of Transkei, Bophuthatswana, Venda and Ciskei. Of these 1 270 persons, 121 are Ministers of government (approximately one out of ten) and in addition there are at least 21 Deputy Ministers. The Central Parliament has 33 Ministers, 21 Cabinet Ministers and 12 Ministers of 'Own Affairs': the 'non-independent black states' have 45 ministers and the 'independent black states' have 43 ministers.

Each of the legislative organs has government departmental structures which, by August 1986, had spawned 151 Government Departments in South Africa. These departments included 18 Departments of Health or Health and Welfare; 14 Departments of Education (under a variety of different names); 14 Departments of Finance or the budget; 14 Departments of Agriculture or Agriculture and Forestry; 12 Departments of Works and Housing; 13 Departments of Urban Affairs or Local Government; 9 Departments of Economic Affairs or Trade and Industry, as well as 5 Departments each of Foreign Affairs, Transport, Post and Telegraphs, Labour and Manpower, Law and Order or Police, Defence or National Security, 3 Departments of Justice, 1 Department of Mineral and Energy Affairs and 1 Department of Environmental Affairs and Tourism. Finally these 140 Departments were responsible to the Departments run by the 11 Presidents, or Chief Ministers, or Prime Ministers in South Africa.

This Legislative network of 3 Houses of Parliament and 10 Legislative Assemblies, with 1 270 members, with 121 ministers and 151 government departments is not cheap to run. [18]

Turning from the national and regional levels of government we now briefly discuss local government. Most of the government's recent efforts revolved around the introduction of African local government. Its previous efforts had been self-created disasters. The Black Local Authorities Act of 1982 introduced village and town councils which closely resembled those for whites. The government hoped to make these councils autonomous by 'suitable financial stipulations'. [19] From the start these bodies faced major obstacles. First, they were presented by government as being a quid pro quo for the exclusion of Africans from the Constitution of 1983. Second, they were introduced without any viable revenue base. Sorghum beer and liquor sales in location

Finally, there has been the unravelling of the administrative system of apartheid. This has been most noticeable in the collapse of the administration of the pass laws and the rapid growth of an informal sector in the black townships. For a long time, bureaucrats within the Administration Boards and other government agencies insisted on strict compliance with housing standards and the regulation of hawking, trading and other black economic activities. The rise of an informal sector, which is estimated at between 25 and 35 per cent of the overall economy, is due firstly to rapid African urbanization which has led to a mushrooming of shacks in backyards and the growth of sprawling shanty-towns on the perimeters of townships. Another factor is the unemployment crisis. Since the early 1970s, the formal economy has been unable to generate sufficient jobs to match the population growth.

Much of the growth of the informal sector has been fuelled by illegal activities. A report in *The Argus* of 22 October 1988, speaks of the rise of a massive shadow economy running on drugs, prostitution, gambling, the illegal sale of liquor, pirate taxis and theft. Some idea of its size can be grasped from official statistics on the confiscation of marijuana (dagga). Dagga seized during 1986 was valued at R2,1 billion and in 1987, at R1,2 billion. According to narcotics experts, these figures do not reflect a drop in use or cultivation, but rather a police force less able to cope with the problem. If the total seized in 1986 represents only half of what is grown per year, which seems unlikely, it means an annual cash inflow of at least R2 billion into the shadow economy. Comparatively, in the formal sector, the civil engineering industry in 1987 received contracts worth about R3,5 billion. All the indications point to the further growth of the informal sector.

Reform-apartheid's ideological dimension : 'sharing' power without losing control

Under classic apartheid the South African government unequivocally favoured whites over blacks. Nationalist politicians articulated the interests of whites and acted on their behalf. In contrast, the essential features of reform-apartheid lie in the government's attempt to present its rule as being in the interests of all South Africans. While government has a clear desire to give all groups a say in government (in Afrikaans the untranslatable term *inspraak* is used), it equally has no inclination to abandon having the final say. In fact, despite the government's protestations, power has been further entrenched in central government's hands and its sense of accountability has, if anything, diminished. Under the Constitution of 1983 (implemented with the

Tricameral Parliament in 1984) the Indian and coloured Houses enjoy no effective veto and the government's apparent attempts to restructure the second and third tiers of government have resulted in a further concentration of power in Pretoria's hands.

There is thus a distinct double-edged quality in the constitutional dimension of reform-apartheid. On the one hand there is a new willingness on the part of Afrikaner leaders to accept that wealth, opportunities and even decision-making, will have to be shared among all the peoples. On the other hand, there is the Afrikaner government's firm conviction that only continued Nationalist rule can ensure progress along that road. Or to put it differently, the government accepts that economic growth, training, job creation and stable food prices are primary goals, as long as these are based on the maintenance of firm control by an Afrikaner leadership. The Nationalist government believes that only Afrikaner leadership can enjoy majority white support and, at the same time, direct all the country's technocratic skills and abilities.

Although the cabinet wants to improve efficiency in its managerial style of government by obtaining 'inputs' from black élites on all levels of government, it refuses to relinquish its grip on any of the levers of political power. This is reflected in a private statement by a prominent business leader: 'We Afrikaners', he said, 'must try to find the secret of sharing power without losing control'.

Power-sharing in any generally accepted sense of the word is anathema to the government at this stage. There are specific Afrikaner fears. As Botha remarked to his biographer: 'The Nationalist Afrikaner has resolved never to be subordinate again in his own country'.[21] Some years ago the Afrikaans writer, N. P. van Wyk Louw, expressed it in even stronger terms when he observed that if the Afrikaners were to become a political minority again, they would be 'as helpless as the Jews were in Germany'.[22] This is not a fear of physical extermination or persecution of Afrikaner leaders[23] under a black government, but rather a fear of a precipitous loss of the political status of the Afrikaners, as was the case in the years after the Anglo-Boer War (1899 – 1902). During these years Afrikaners found themselves squeezed between British capital and the black working class and lost control over their schools.

There are also concerns over the vulnerable position of white civil servants and farmers under a future black government. In terms of the 60 : 40 demographic ratio in white society, the Afrikaners are not over-represented in the civil service, but they do dominate the upper

levels. It is quite conceivable that one of the first objectives of a black majority government would be to Africanize these levels of the civil service. It is also likely that a radical black government would redistribute land and aid in favour of African farmers.

In the course of the 1980s the government has begun to de-emphasize its concerns about Afrikaner needs and interests. It now stands as the instrument of white communal political power. Since whites control more than 90 per cent of the instruments of production (land and capital) it is strongly opposed to any radical redistribution of income or assets. Instead it proposes that blacks abandon radical strategies in favour of working within existing structures to address income inequalities.

In thinking about 'power sharing', the government has firmly ruled out the Westminister style of democracy whereby the winner takes everything and the minority party is effectively excluded from power, at least until it wins an election. The government has also ruled out a class-based constitutional structure (for instance, a qualified franchise) which may result in multi-racial middle-class rule.

Instead it intends to reform the constitution slowly and deliberately, pausing at each stage to give the white electorate a chance to accustom itself to the changes. Despite the disastrous alienation of Africans which the Constitution of 1983 produced, the government still believes it had no alternative but to co-opt Indians and coloureds into the parliamentary system, while leaving Africans out. As a second phase, the government may attempt to incorporate Africans in the executive up to cabinet level, merge the existing three Houses of Parliament (while still insisting on separate ethnic voting under a system of concurrent majorities) and introduce a separate African chamber. However, for the foreseeable future, the government will continue to insist that whites elect their own representatives for the national level of government. The reason for this demand is simply that in an election based on the system of proportional representation the National Party could well be reduced to insignificance. Ruling political parties are not in the habit of legislating themselves permanently out of power.

The suggestion has often been made that the State President may, at some stage, want to rid himself of all the party constraints, or that the military may become too impatient with the slow rate of change, or that one or a combination of these factors may produce the suspension of Parliament and the introduction of rule by the civilian executive and military. Some observers have concluded that this country is already effectively ruled by the State Security Council (SSC) which is chaired by the State President and comprises the inner cabinet and the country's main security officers.

In our view this interpretation does not square with the facts and underrates some important factors. First, there is the fact that the Afrikaner nationalist movement is split between the National Party and the Conservative Party. This means that it would be extremely difficult for the government to suspend the constitution. If it were to do so, its legitimacy and security problems would be compounded by a vengeful white right wing claiming that it had been robbed of an inevitable victory. Another important factor is the fact that Nationalist politicians have always jealously guarded their position of political authority and would be most reluctant to accept generals and brigadiers as political actors. It is undoubtedly true that some military officers have displayed considerable 'private initiative' in South Africa's destabilization of neighbouring states. They could not have done this, however, had those in the highest positions of political authority not been prepared to turn a blind eye.

Many observers have been fascinated by the growth of the National Security Management System (NSMS) positioned below the SSC and parallel to the civilian structures of government (see Figure 4.1). According to Annette Seegers, an academic authority on the subject, the system is comprised of the following elements.[24] Firstly, to co-ordinate the activities of state departments, particularly with respect to the administration of blacks, Joint Management Centres (JMCs) were created for each of the twelve official regions of South Africa. Within each official region there are approximately sixty sub-regions, and for each, a Sub-Joint Management Centre (Sub-JMC) was created. A Mini-Joint Management Centre (Mini-JMC) was introduced for each of the 448 mini-regions into which the sub-regions were divided. Finally, Local Management Centres (LMCs) co-ordinate state functions on municipal level.

Four committees exist at every level of the NSMS: one deals with security, another with constitutional, economic and social affairs, a third with communication and local propaganda, while the fourth one draws together the chairmen of the other committees. The NSMS bodies are expected to liaise with the local community. They are encouraged to establish liaison forums with developmental associations, which also exist on regional, sub-regional and mini-regional, and municipal levels. At this level, state officials are brought together with community leaders, businessmen, religious leaders and educators.

These structures have several aims. Firstly, they serve as an early warning system of any build-up of the 'revolutionary' climate in the townships. Secondly, they channel security briefings from the top down to the lowest level of the bureaucracy and to local community leaders working within the parliamentary system. Thirdly, the state uses these structures to bypass bureaucratic bottle-necks in order to redress grievances promptly and to establish the material needs of the black community. Fourthly, they are used to wage a propaganda war aimed at thwarting the activities of extra-parliamentary groups.

Because officers from the South African Defence Force play a prominent role in the SSC and on the lower levels of the NSMS, some observers have gained the impression that these structures enable the military to 'rule' South Africa.

Some fairly obvious comments need to be made. Firstly, the NSMS is under the overall control of a civilian politician who has no security background. The particular function of the NSMS seems to be the efficient co-ordination of sluggish and poorly co-ordinated civilian bureaucracies. Secondly, the NSMS on all its various levels is sustained by a crisis, such as the one South Africa experienced in 1984 – 6. Once the crisis diminishes, the importance of the NSMS decreases proportionally and bureaucratic inertia sets in. Thirdly, owing to limited financial resources, the NSMS can only concentrate on the worst trouble-spots; it lacks the means to become an effective alternative system of administration throughout the country. It is in the realization that civilian administration is ultimately cheaper, that the state spent so much time and energy in promoting local government and the elections of October 1988.

Finally, the significance of the State Security Council must be seen in perspective. It is this body that makes crucial decisions, such as calling a State of Emergency in 1985 and 1986, with the full cabinet apparently acting as a mere rubber-stamp. However, the key fact is that the SSC is a body in which civilian politicians take responsibility for decisions. Furthermore the body was designed to suit the governing style of P. W. Botha who has had close ties with the military since the mid-1960s and who preferred force and socio-economic upgrading to negotiation as a means to achieve stability. There are signs that F. W. de Klerk, Botha's successor, is keen to diminish the considerable influence which the security forces managed to gain during the 1984 – 6 unrest.

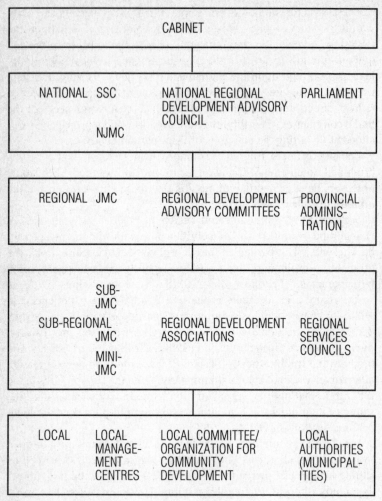

Figure 4.1 National Security Management System

Source: Annette Seegers, 'The National Security Management System: A Description and Theoretical Inquiry', paper presented to Africa Seminar, University of Cape Town, 1988.

The government's social reforms : grafting reform-apartheid upon apartheid

The biographer of P. W. Botha remarks candidly in his perceptive study: 'In short, it is for Botha impossible to dismantle the entire structure of apartheid and he is not planning to do so'. Under Botha the government started to graft reform-apartheid upon apartheid. It eliminated some of the negative or outdated features such as state-backed privileges for whites, statutory job reservation for whites, the racial salary gap in the civil service, and unilateral white decision-making without any consultation or persuasion. What is also new is the government's promise to move, over an unspecified period, towards the goals of equalized spending on social services for whites and blacks, equal opportunities for everyone, equality before the law, common participation in decision-making in pursuit of the democratic ideal, and full human rights for everyone.

These are momentous principles, but what is missing, as analysts such as André du Toit have argued, is a framework and a recognizable goal which would give coherence and structure to the many and various changes in government policy. In the absence of any clear and distinctive view of the political goal, these abstract and general principles have to be interpreted in terms of the government's known positions at the lower levels of policy, rather than the other way round.

Thus, the principled rejection of racial discrimination must be squared with the government's insistence on segregated education, segregated residential areas, and the classification of the entire population according to racial categories. The commitment to open-ended negotiations must be reconciled with the rejection of negotiations with the ANC, and the prolonged imprisonment of political prisoners, such as Nelson Mandela and other ANC leaders, as well as the outlawing of the ANC itself. The government's commitment to democracy must be coupled with the rejection of any system other than one based on racial classification and in which representatives of the various racial groups will 'share power' on the executive level.

This tendency of heavily qualifying reformist intentions with below-the-line policy commitments has produced confusion and outrage among those who deal with the South African government. Some call it double-talk, others the shifty vocabulary of a foreign culture. The Botha government is deeply pained by these accusations. The reason for the confusion is obvious. When the Botha government talks about the removal of racial discrimination, its critics understand this to mean that it intends to abolish segregated education. For the Botha government, however, this means that state spending on education for the four racial groups must be equalized over a certain period. For government it then becomes a question of finding the resources and the right formulas in order to bring about equality in state spending on education, and establish equal standards, which means adjusting white standards downwards. In practice, the resources are rarely available, the formulas are usually contorted and whites soon rally against the lowering of their standards.

In the same way, the government rejects negotiation with the ANC on the grounds that no Western government negotiates with terrorists; it then goes out of its way to 'prove' that the ANC is a terrorist organization. While these arguments may satisfy the government's own constituency, it does not solve the problem of finding credible black leaders with whom to negotiate.

The key difference between a coercive system and a power-sharing system lies in the significance of bargaining. Hard bargaining — that is, when a bargain is haggled, struck and kept — is a concrete sign that power-sharing is operating successfully. In South Africa the hard bargaining has hardly begun; the country has just entered the pre-negotiating phase. Real negotiation would entail hard bargaining about the fundamental principle of freedom of association.

For the moment the NP is still quite explicit in its insistence on segregated schools, hospitals, residential areas and political institutions. In the words of F.W. de Klerk, during the opening session of the 1986 Parliament :

> The National Party believes the following. Firstly the recognition of the importance of group existence is not discriminatory *per se,* in fact it is a condition for peaceful existence... Secondly there are certain fundamental matters which are inextricably attached to group security. The most important of these is that each nation, each group should have a community life of its own. This entails residential areas of its own, schools of its own, institutions and systems of its own within which the group is able to preserve its own character and look after its own interests... [It] means that there should also be certainty in regard to the definition of each group... Moreover, it means that each group must have a power-basis of its own, a power-base within which the group is able to take care of its own affairs by itself.[26]

There is little in this exposition of the 'autonomy' of state-defined racial groups with which an ideologue of classical apartheid would quibble. De Klerk went on to add that it is from this group 'power-base that group leaders could co-operate, share power, make joint decisions with the leaders of other groups on matters of common interest'.[27] De Klerk has thus, almost without pausing for breath, grafted reform-apartheid upon apartheid.

The functioning of reform-apartheid

The four main institutions of reform-apartheid are the National Party, Parliament, Regional Services Councils and the homelands.

The National Party

The NP is undoubtedly the central institution of the entire constitutional structure of reform-apartheid. Should it for some reason or other lose power, the constitution would collapse with it.

The transition from classic apartheid to reform-apartheid has had major implications for the NP. Up until the mid-1970s it was able to appeal to the interests of all the classes within the Afrikaner ethnic group. It has been estimated that, in the general election of 1977, more than 85 per cent of the Afrikaners voted for the NP. Since then the situation has changed considerably. White workers have seen the traditional protection they enjoyed being withdrawn as a result of the labour reforms. Farmers have found that the state's priority has shifted away from giving aid to white farmers to cutting farmers' subsidies and ensuring stable food prices in the cities. Lower level civil servants have experienced much less government sympathy for their salary demands. All white classes, except the very rich, have seen their income eaten away by high inflation and a sagging economy. Sanctions and disinvestment have aggravated the economic problems. Chris van Wyk, Managing Director of Trust Bank, estimated that by 1990 the average South African could, over the medium to long term, be a third poorer as a result of sanctions and disinvestment.

Because of these developments, Afrikaner support for the National Party had dropped to approximately 52 per cent by June 1989. Nevertheless it has continued to draw support from across the income spectrum. If only Afrikaners are taken into consideration, roughly 40 per cent of the NP support comes from the high income group; 30 per cent from the middle income group and 25 per cent from the low income group. However, if the white population as a whole is considered, the NP is supported almost equally by the three income groups. It follows that English-speaking blue-collar workers are more prepared to support the government than their Afrikaner counterparts. This means that the NP has to juggle class interests constantly and cannot act expressly on behalf of any one class or ethnic group.

Although the NP constantly keeps the right wing in mind when assessing the pace of reform, this factor can be exaggerated. It is definitely not only fear of the right wing that has prevented the NP from making more progress with dismantling other aspects of political and social apartheid. Most members of the cabinet believe that apartheid's 'group basis' is a prerequisite for stability. They are convinced that once the groups are no longer defined by government, and that once all the barriers to a mixed society are relaxed, conflict will become uncontainable.

Parliament

The short-comings, defects and the general negative impact of Parliament under the new constitution have been enumerated so often that they need only be briefly stated here. The introduction of the Tricameral Parliament has made it far more difficult to resolve the issue of African political rights. Surveys have shown that Africans are more opposed to this constitution than they ever were when whites ruled alone. The new Parliament has also been unable to demonstrate to the Indian or coloured electorate that the new system has made any significant impact on the vital issues of jobs, housing and, above all, education.

The most serious aspect, however, is the fact that apartheid has been constitutionalized. The constitution allocates control over education, health and 'group areas' to the relevant 'own affairs' authority. In other words, coloured and Indian members of Parliament now control their 'own' group areas, education and health services. This has provided more power and patronage to coloured and Indian politicians and a greater number of civil service jobs to members of these communities. As a result, a small proportion of both the coloured and Indian populations have been given a vested interest in the new constitution.

All of this has, at the same time, greatly increased inter-group conflict in Cape Town and Durban within the coloured and Indian communities respectively. Most middle-class coloureds and Indians decisively reject the 'own affairs' concept and the whole system of race classification and group areas. In Cape Town, as few as 5 per cent of the potential coloured voters cast their ballots for Parliament in the 1984 election. It follows that a Minister of Education, who is elected on such a ridiculously low poll, would evoke more outrage if he closed down schools than would have been the case had a white minister done so. Another implication of the 'own affairs' policy and practice is that it makes it all but impossible to devolve control over education or health to new regional governments or federal sub-states. For instance, the education of Indians cannot be transferred to a KwaZulu-Natal power-sharing authority because it already falls under the Indian 'own affairs' authority.

Yet it would be wrong to consider all of the effects of the Constitution of 1983 as bad. On a symbolic level it has significantly diluted the white group's sense of superiority and exclusivity. It has helped whites to become accustomed to blacks in cabinet and to future African parliamentary representation. The question is no longer how to exclude Africans, but how to accommodate them in a common system. The Tricameral Parliament has facilitated this process.

Secondly, the presence of coloureds and Indians in Parliament has considerably increased the pressure on government to narrow the gap in social spending and civil service salaries between whites on the one hand, and coloureds, Indians and Africans on the other. It may well be that the coloured people and Indians will give the constitution considerably more support once it no longer rests on the compulsory segregation of schools and residential areas. Thirdly, in Parliament select and standing committees drawn from all three houses are exercising more influence on legislation than was the case under the old system. Before the new constitution was implemented, the NP caucus decided on every aspect of strategy with respect to the various stages of bills passing through the House. In the new system the caucus is more divorced from cabinet and, as a result, there is a greater degree of open-endedness in the NP's parliamentary strategy.

In general, the new constitution has played a small role in preparing whites for multi-racial rule. It has, of course, occurred at great costs. To Africans and to middle-class coloured and Indian people, the constitution and the Tricameral Parliament remain an insult.

Regional Services Councils (RSCs)

The initial motivation for these new structures was fairly complex. The Croeser Committee had made suggestions on the most appropriate methods of financing major services and infrastructure development within local communities. This was a critical issue, especially in the case of the new black town and village councils which, because they were located almost exclusively in dormitory townships, had no tax-base to speak of. Some form of authority at the metropolitan or regional level, capable of raising and transferring revenue, was long overdue.

Other motivations were largely political. The established National Party principle of group self-determination meant that local communities had to be strictly segregated in terms of residential rights and control over community affairs. The appropriate vehicle for the realization of this principle was the municipality. Some means had to be established for maintaining or creating separate municipalities for different groups without depriving the less-centrally situated Indian, coloured and African local areas of the benefits of the tax-base offered by central business districts and industrial areas. A regional authority capable of providing integrated major and expensive services to the member local authorities, which differed markedly in terms of financial viability, was therefore a way of justifying segregation of local authorities.

Subsequent to the announcement of the Regional Services Councils, most of the major Indian and coloured suburbs, which the government had helped to incorporate under RSCs as separate local authorities, have refused to accept this status. Provision has therefore had to be made for membership of the RSCs of the so-called Local Affairs Committees and Management Committees. Provision has also been made for administrations to exercise discretion in admitting other bodies below the level of municipal authorities, hence creating the possibility of informal organizations (such as church community bodies) being represented.

The RSCs are a mixed bag in political terms. On the positive side, the Councils are multi-racial in composition and involve increasing shared decision-making. The Act which established these bodies also states quite explicitly that priority has to be given to the needs of the less-advantaged or less-developed communities in the provision of new services or the upgrading of existing services. The list of functions performed by RSCs is also fairly extensive. Added to this is the fact that the RCSs are able to deliver additional services on an agency basis for any member local authority or area. The scope for the upgrading of services in African, Indian and coloured local areas is therefore very wide indeed. Theoretically, the RSCs have the capacity to improve the quality of life in townships. Whether they will do so in practice still remains to be seen.

These benefits have to be measured against the critical short-comings of the RSCs. They reinforce and legitimize the segregation of municipal areas according to race at precisely the time when demands for multi-racial and fully integrated municipalities are growing. During the 1984 – 6 unrest, existing African local government was effectively challenged in many towns, particularly in the Eastern Cape, by civic associations and other forms of local mobilization. The pressure for multi-racial municipalities was, in some towns, taken to the point at which civic association leaders were negotiating with local chambers of commerce over the composition of an integrated metropolitan authority. Since representatives on RSCs are not directly elected but are nominated by municipal councils, the new bodies depend largely on the credibility of these town and village councils, which were introduced by government, for their legitimacy. In some areas these bodies have come under a lot of pressure. The local councillors are targets of violent attacks by township dissidents whenever any unrest occurs. This form of action has further weakened the position of African local authorities which, from the outset, had poor local support, as evidenced in low election polls.

The weight of representation is a further problem, since local authorities enjoy representation based on their consumption of services. Hence the richer authorities dominate the RSCs, even though there is a fixed upper limit to the size of their representation. African local authorities will continue to have only one or two representatives until such time as their utilization of services improves.

The consequences of such factors, plus the fact that the legislation providing for RSCs was not presented to African leaders for discussion and negotiation prior to its formal announcement, has presented the government with a situation in which significant African leadership has rejected the new councils.

At the same time there has been resistance to the new deal from conservative white councils, which are less than interested in a new structure which gives priority to the upgrading of black areas within their regions. Thus the RSCs, at the third tier of government, have provided evidence anew of the extreme delicacy with which political reform has to be negotiated and introduced. Clearly the government's thinking behind the RSCs, although progressive and sensible in some respects, also involved a serious miscalculation of the reception that the councils were likely to receive.

This range of factors should also be seen in relation to the likely effectiveness of the RSCs in providing an upgrading of services of an order likely to be expected of them. Two financial problems in particular bear consideration.

Firstly, the RSCs are expected, at some stage, to take over the provision of transport subsidies currently funded by the central treasury. These sums are of a magnitude which would absorb a substantial proportion of the finance presently raised by these RSCs. This means that either commuter charges will increase substantially, which would be politically hazardous, or the RSCs' ability to provide other services will be severely curtailed.

RSCs are financed from taxes which are levied on all businesses in a region, and are based on turnover and wages. There have been considerable misgivings about this within industrial circles and it is generally felt that the new taxation will inhibit the business expansion in metropolitan areas or, at the very least, prove to be an additional incentive for decentralization to rural areas. If the latter occurs, unemployment in the metropolitan areas will increase and, as there is already an excess of labour at the decentralized growth points, the unemployed urbanites are unlikely to be attracted to these areas.

Homelands under reform-apartheid

It is clear that the government no longer envisages the homelands as an alternative to political rights in the common area of the RSA. A process of re-incorporation of these areas into the central political dispensation has begun. At economic level this process is even further advanced. This is evident from the fact that the economic planning regions, as defined by the Department of Constitutional Development and Planning, extend across political boundaries, as do the Regional Services Council boundaries in metropolitan areas abutting certain homelands. The National Party's observer status at the KwaZulu-Natal Indaba was a further pointer to the rapid revision taking place in the government's view of the future of the homelands. It is now commonly accepted in government circles that the policy of separate development has failed in the political functions originally intended for it by Verwoerdian planners.

A few points need to be made regarding the internal characteristics of these regional structures. Firstly, there can be little doubt that the following has occurred over time:

- Considerable vested interests have developed within the homeland bureaucracies and among their élites.
- A great deal of administrative and political experience has been acquired by Africans within these areas.
- The level of legitimacy and support for these regional governments is substantially higher among populations within the homelands than it is, for example, in the third tier of government for blacks within the common area.
- The homeland security services, both formal and informal, have managed to enforce 'law and order' much more successfully than has been the case in common area townships. It is a popular argument that this has been achieved through naked coercion and terror. While the homeland security services have not been gentle by any stretch of the imagination, the same can be said of the South

African security services. Unless one makes the unlikely assumption that the rather poorly staffed homeland security services are somehow more efficient than their South African equivalents, one has to reach the conclusion that the relative stability in the homelands rests on a more comprehensive range of factors than is the case in the common area. In 1985 – 6 the security machinery in the homelands was tested very severely in the northern and eastern Transvaal, in Lebowa, Gazankulu and KaNgwane, and in Bophuthatswana. However, the social order proved to be resilient.

• Finally, the growth rates in the homelands, while falling far short of the requirements of economic viability in the short to medium term, have been much higher than in the RSA as a whole, albeit from low base-lines. The homeland administrations have also been able to engage in more meaningful experiments in development than has been the case within the more established institutional constraints in the common area.

The overall pattern, then, is that the policy of separate development, and of homelands in particular, is completely discredited among Africans in the RSA common area. However, the situation is more complex among the *de facto* homeland populations. Here the grafting of modern administrations upon tribal authority structures and the emergent industries are generating a small degree of legitimacy. The overall economic, political and administrative development has been such that structures now exist that have some internal viability in terms of decentralized regional government.

The homelands seem destined to lose their original functions as politically autonomous areas, but may acquire a new and more viable status in a future process of power-sharing in South Africa, if the boundaries of such regions become modified so as to enable them to lose their tribal connotations. At the same time they are hardly any threat to the scope for mobilization by black political movements within the massive urban constituencies of the common area.

Conclusion

This chapter has argued that some major economic, demographic and political forces have propelled the process of change, and that the government's initiatives and responses are inadequate and are often futile attempts to exert some control over the pace and direction of change. Instead of being directed to an overall goal, attempts at reform have been rather *ad hoc* and unco-ordinated. Stripped to its basics, the reform process originated as a result of the shortage of skilled labour which the economy began to experience in the late 1960s. This, together with the fact that the reformed policy was not supposed to be discriminatory, forced the government in the early 1970s to expand black education and training. However, the government did not pro-actively try to adapt the political system. Instead the government thinking was, in P. W. Botha's words: 'blacks had to be uplifted and the consequences of this will have to be accepted'.[28]

The government has painfully learnt how dangerous it is to 'uplift' people through expanded education and training, without also modifying the political system or being able to rely on a strong economy to absorb the newly educated. It is indeed ironical that the apartheid state embarked on a massive expansion of black education and training in the early 1970s in the expectation of continued rapid economic growth, when it was at this point that the South African economy entered a period of prolonged stagnation and depression from which it has not yet recovered. The total manufacturing production figure for 1988 was not as high as it was in 1981, while real domestic economic activity in the first quarter of 1988 was only 6 per cent above the level of the fourth quarter of 1981. In 1986 some 147 000 fewer people were employed in the formal economy than in 1981. Black children who have educational qualifications of which their parents could only dream and who have much higher social and political aspirations, find themselves without employment or in a jobs for which they are over-qualified.

The experience of Lebanon, which also rapidly expanded education in the decade prior to the outbreak of civil war, shows that the sudden increase in the number of educated unemployed can be the recipe for political disaster. The South African state is, of course, much more powerful than Lebanon, but its modernization programme has also unleashed formidable forces.[29] Although black aspirations have been raised, these aspirations have not been fulfilled. Blacks are unable to compete with whites in the job market and are unable to attain equal status on the basis of educational qualifications.

In noting the upswing in the destabilizing effects of reform and pointing to the surge of support for the Conservative Party, cabinet ministers have privately stated that the government may have reached the limits of the degree of reform acceptable to the white constituency. Yet merely maintaining the status quo in South Africa is not really an option. In the first place, South Africa, unlike Northern Ireland and Israel, is ruled by a small and shrinking minority, not a majority controlling a society with a fairly set population ratio.

In the second place, South Africa is a developing country. It has considerable unrealized economic potential. To achieve optimal growth, South Africa needs capital and technology from Europe and the United States of America. Instead its failure to proceed fast enough with real reform has led to the introduction of sanctions by the West which, in turn, has compelled South Africa to become an exporter of capital. Only genuine political reform will allow South Africa to regain access to financial markets abroad.

Lastly, the state of the economy is vastly more important in South Africa than it is in either Israel or Northern Ireland, which enjoy virtually guaranteed injections of capital from Washington and London respectively. In South Africa, even the security forces fear a stagnant economy, since spiralling unemployment heightens the security risk. Businessmen regularly tell the government that unless some form of political accommodation is reached shortly, the private sector will lack the necessary confidence to make further new fixed investments.

In the case of Northern Ireland it might make sense to settle for an acceptable level of violence. In South Africa, however, urban violence is potentially far more explosive and massive than in Northern Ireland. Hence, merely maintaining the status quo is not a viable option. For South Africa to realize its true potential, there is no alternative but to proceed beyond reform-apartheid to the non-racial restructuring of society. If we focus only on the political aspects of apartheid (i.e. post-1948 policies) the following are the most important remaining policies which will have to be changed:

- the registration and the division of the people according to the Population Registration Act;
- the exclusion of the African population from equal political rights at the level of central government, where virtually all power is concentrated;
- the maintenance of segregated residential areas, state schools and hospitals, and some of the choice public beaches and resorts.

However, if we consider the entire segregation order which sprung up after the opening of the diamond- and gold-fields in the 1870s and 1880s, the following are the worst remaining elements:

- the migrant labour system and, in particular, the compound system on the gold-mines;
- the discriminatory spending on education and other welfare services which results in unequal opportunities for the various groups throughout life;
- the Land Act which prohibits black land ownership in four-fifths

of South Africa;
- the near exclusion of blacks, and particularly Africans, from the top levels of central government, the executive and managerial levels of parastatals and private companies;
- the lack of black control over the means of production (mines, factories, etc.).

Even if the government did have the will to embark on restructuring in these cases, it would require far better directed pressure and a much broader coalition of forces than the elements which produced reform-apartheid during the 1970s and 1980s.

Notes

1. Pierre L. van den Berghe, 'Protection of ethnic minorities' in Robert G. Wirsing (ed.), *Protection of Ethnic Minorities — Comparative Perspectives*. New York: Pergamon, 1981, p. 345.
2. The unravelling of the system and the bureaucratic tangle is a main theme of Stanley Greenberg, *Legitimating the Illegitimate: State, Markets and Resistance in South Africa*. Berkeley: University of California Press, 1986.
3. Dirk and Johanna de Villiers, *PW*. Cape Town: Tafelberg, 1984, p. 89.
4. Merle Lipton, 'South Africa : authoritarian reform?', *World Today*, 1974, p. 248.
5. *HAD*. 1973, col. 346.
6. Phil Weber, *Republiek en Nasionale Eenheid*. Stellenbosch: University of Stellenbosch, 1973, p. 13.
7. M. Holden and M. McGrath, 'Economic Outlook', *Indicator South Africa*, 1985.
8. Nedbank Group, *Economic Unit Guide 1985*. M. O. Sutcliffe, 'The Crisis in South Africa: material conditions and reforms and response', unpublished paper, 1987.
9. *Business Day*, 10 July 1986; *The Star*, 2 December 1985, p.21; *Financial Mail*, 2 October 1987; *Finansies en Tegniek*, 23 September 1988, p. 21.
10. B. M. Schoeman, *Vorster se 1000 Dae*. Cape Town: Human and Rousseau, 1974, p. 20.
11. An illuminating analysis of the tortuous adaptation to the sports policy is by William A. Munro, 'The state and sports: political maneuvering in the civil order', Wilmost G. James, (ed.), *The State of Apartheid*. Boulder: Lynne Rienner, 1987, pp. 117 – 42.
12. Merle Lipton, *Capitalism and Apartheid: South Africa, 1910 – 1986*. Aldershot: Wildwood House, 1986, p. 59.
13. N. E. Wiehahn, 'Industrial relations in South Africa — a changing scene', in D. J. van Vuuren, et al., (eds), *Change in South Africa*. Durban: Butterworths, 1983, pp. 189 – 90.
14. Lipton, *Capitalism and Apartheid*, pp. 66 – 7.
15. Simon Brand cited by Lipton, *Capitalism and Apartheid*, p. 59.
16. *HAD*, 1978, col. 579.
17. *Race Relations Surveys, 1985*. Johannesburg: SA Institute of Race Relations, 1986, p. 59.
18. Michael Savage, *The Cost of Apartheid*. Cape Town: University of Cape Town, 1986, p. 8.
19. *HAD*, 1982, col. 9623.
20. This is based on work being currently done by Gerd Behrens on the 'other two Houses of Parliament' for a University of Cape Town doctoral dissertation.

See also *The Argus*, 21 August 1987; *HAD*, 1987, House of Representatives, cols. 121 – 30.

21. De Villiers, *PW*, p. 34.
22. N. P. van Wyk Louw, *Liberale Nasionalisme*, 1958, reprinted in *Versamelde Prosa I*. Cape Town: Tafelberg, 1986, pp. 502 – 6.
23. This fear cannot be dismissed entirely. A joke is currently circulating in Afrikaner circles of Botha in 1990 insisting that R10 billion be voted for prisons but only R10 million for education. When challenged he asks: 'Do you for one moment think the blacks will put us in school?'
24. The paragraphs of the functioning on the NSMS is based on Annette Seegers, 'The National Security Management System: A Description and Theoretical Inquiry', a paper presented to Africa Seminar, University of Cape Town, 5 October 1988.
25. De Villiers, *PW*, p. 154.
26. Debates of the House of Assembly, *Hansard*, 1986, no. 1, pp. 145 – 6.
27. *Hansard*, 1986, no. 1, col. 146.
28. De Villiers, *PW*, p. 90.
29. See the comparison of Lebanon and South Africa by Theo Hanf, 'Lessons which are never learnt', *South Africa – a Chance for Liberalism*, Friedrich Naumann Foundation, (ed.). Sankt Augustin: Liberal Veralg, 1985, pp. 339 – 54.

PART TWO

Towards coexistence

Towards coexistence

In Chapter 4 we reviewed the major institutional factors which caused the policy mutation which we termed reform-apartheid. The process is far from complete. Not only are major forms of apartheid still deeply entrenched, but the pressures for further change within the society have heightened significantly, resulting in greater polarization.

There is a common perception, both nationally and internationally, that the present phase of reform-apartheid will quickly pass. Although South Africa is widely depicted as being in a situation of stalemate or deadlock there is also the perception that things are about to change.

There is an assumption that the major economic, administrative and political pressures reviewed in Chapter 4 will inevitably shift South Africa beyond the current phase into the post-apartheid era. A plethora of conferences and symposiums, as well as titles of essays and papers, seem to signal the death of apartheid and the inevitable dawning of a new order free from apartheid.

In this regard, thoughts about the resolution of conflict in South Africa are frequently very clear and uncluttered. Speaking in 1986, Mr Genscher, the West German foreign minister, was remarkably unconcerned about the complexities of the South African power equation when he stated boldly that 'what is needed is the immediate dismantling of apartheid and the establishment of equal civil rights for South Africans, including the "one person, one vote" principle'.

In an address to the Royal Commonwealth Society in London in 1988, Sir Geoffrey Howe, the British Foreign Secretary, was equally confident when he repeated an appeal which has become virtually the standard international position on South Africa. The South African government and genuine, black South African leaders, among whom he included ANC officials, should *set aside* present strategic commitments and negotiate a resolution of the South African conflict.

In the final analysis, no responsible South African can afford to dismiss sentiments such as these. It is interesting to note that the various polls conducted over the years have shown that even among those whites who would not specifically endorse negotiations with the ANC at present, there is an underlying realization that, sooner or later, an agreement has to be struck with all representative black leaders.

More recently, for example, in a poll conducted in October 1985 by Market and Opinion Surveys, 70 per cent of all whites, 59 per cent of Afrikaners and even as many as 41 per cent of Conservative Party supporters indicated that they considered it inevitable that whites would ultimately have to negotiate with the black majority on power-sharing. In many quarters there is an unquestioning acceptance that the knot of the conflict in South Africa will never be completely untied unless the black majority in the country is ensconced in its rightful position of power and influence. We fully share this perspective.

The problem with assessments which work backwards from the ultimate resolution, however, is that they create the impression that a society is nothing more than an elaborate moral Meccano set and that all that is required is a simple decision to dismantle one construction and replace it with another.

We believe that there is no simple solution. The prospects for the type of break-through conjured up by hopeful external observers are slim indeed. As indicated in the conclusion to Chapter 4, we accept that there are processes at work which are constantly shifting, pushing, adjusting and realigning many of the key elements in the present order, but the notion of a swift transformation of society is over-ambitious and simplistic. Scope certainly exists for a variety of strategies and approaches which will culminate in a transformed South Africa. Realizing these effects, however, will take time, commitment and, above all perhaps, sound analysis of the situation.

In the next chapter we examine the two most widely discussed of these grand solutions, as well as the major strategies for change which have commanded the attention and energy of both observers and activists alike. We then contrast these strategies with changes which have occurred as a result of effective pressure. This is followed by a brief forecast of likely changes over the next five years. Against this background we put forward a framework for mutual accommodation.

We conclude with a discussion of those approaches which are likely to be the most telling and effective in the resolution of South Africa's conflict.

CHAPTER 5

Grand solutions and major change strategies

In this chapter we analyse the most commonly discussed grand solutions and major change strategies suggested for South Africa. These include the common society model, first put forward by liberals and recently also by some progressive organizations in the extra-parliamentary movement, and the 'nations of minorities' theory propagated by government. We argue that neither of these solutions take into account the hard realities of South African society. Neither is likely to win decisive support from both sides of the potential divide so neither can be regarded as a solution in any real sense of the word.

The common society model

The ideal of a common society is almost an inevitable response to the manifestly dire implications of the underlying South African conflict. The ideals of unity, non-racialism and collective empathy are, in a sense, also a direct and common-sense reaction to apartheid, and have been for decades. These ideals have been codified in many charters and programmes, and today stand as the goals of most opposition groups left of government.

How realizable are these goals? Do they facilitate or impede a solution in South Africa?

The loss of support which the Progressive Federal Party (PFP) suffered in the General Election of 1987 and the party's subsequent act of self-immolation have brought us to the end of an era. In the decade before 1987, the PFP sought to win sufficient electoral backing to enable it to participate in the 'politics of power' rather than the 'politics of protest'.

The PFP championed the idea of liberal democracy, which was based on individual participation in a multi-party, competitive system albeit within some form of geographic federalism. It rejected representation of defined racial or ethnic groups. The party argued that increased support for this non-racial model would be the only effective means of demonstrating to the black majority and to the international community that increasing numbers of whites were prepared to break decisively with apartheid.

In retrospect it is clear that the PFP approach had no prospect of winning strong electoral support. Opinion polls, conducted by Market and Opinion Surveys in the period 1974 to 1987, consistently showed minimal white support for a system in which they would no longer be in a position to choose their own (white) representatives.

Table 5.1 shows the results of a poll taken in October 1988, ten years after the PFP had started to propagate its policy with the support of virtually all the English-medium newspapers.

TABLE 5.1: PREFERENCES FOR A POLITICAL-CONSTITUTIONAL MODEL AMONG WHITES IN THE PRETORIA-WITWATERSRAND AREA
(Sample: 499 whites, stratified random selection)

	Total (n 499)	English-speakers (n 267)	Afrikaners (n 239)
Parliament should return to a whites-only House of Assembly	21	10	34
Parliament stay as it is	13	6	20
Introduce fourth chamber for Africans	25	30	19
Single mixed Parliament with majority in control	7	11	3
Parliament become an umbrella body for regions representing all groups	31	40	21
No answer	3	3	3
Total	100	100	100

Source: Market and Media Research, Johannesburg, for Argus Group Co., October 1988.

The PFP policy did attempt to safeguard the white minority position through its federal provisions and minority veto. However, nothing in its policy guaranteed *participation* of minority parties in the executive arm of government. Furthermore, its political communication suggested unqualified democratic idealism. In this sense the position in Table 5.1 coming closest to PFP policy is the 'single mixed Parliament with majority in control' option which received only 7 per cent overall support.

In the same study roughly similar support was recorded for the proposition that 'government should change immediately to a non-racial system'. It is interesting to note that the percentages favouring Nelson Mandela in power were: 2 per cent of the total, 3 per cent of English-speaking whites and less than 1 per cent of Afrikaners, as opposed to 27, 39, 12 per cent respectively in favour of Buthelezi as leader.

In another study, undertaken in 1986 and repeated in 1989, the following responses were recorded.

TABLE 5.2: PREFERENCES FOR A POLITICAL DISPENSATION (IN PERCENTAGES).
(1986 sample: 1804 whites nationwide—stratified random sample; 1989 sample : 1504 whites ; 1989 results in brackets).

	Total (n 1804)	English-Speakers (n 755)	Afrikaners (n 1048)
Separate freedoms with self-determination in own homelands or core areas	20 (19)	10 (6)	28 (27)
Unitary state, one Parliament, one person one vote	5 (4)	8 (7)	3 (2)
Power-sharing on an ethnic group-basis with group autonomy	27 (27)	19 (19)	32 (32)
Negotiations between all groups on a non-ethnic regional basis with rights and privileges of individuals guaranteed	37 (41)	53 (59)	25 (28)
None of the above / don't know	11 (11)	10 (10)	12 (12)
Total	100 (100)	100(100)	100 (100)

Source: Market and Opinion Research (Pty.) Ltd, Bellville, Cape, November 1986 and May 1989.

Table 5.2 reveals that the 'unitary state, one person one vote principle', which comes closest to the demands of external politicians such as Foreign Minister Genscher and to the non-racial ideal, is supported by less than 10 per cent of whites. The fourth option receives substantial support due to the fact that 'group' negotiation is involved and the regional 'federation' would limit the power of the majority.

The losses suffered by the PFP in the 1987 election indicated that the electorate considered the proposed jump from apartheid to the non-racial model as far too big. Whites may be prepared to move toward such an outcome in stages, but they are not prepared to accept a sweeping transformation over which they have no control.

The 1987 election left whites left of government with a huge political hangover. Within the PFP there were few who were prepared to go back to the politics of protest of the early 1970s, when they had only Helen Suzman in Parliament to campaign for the liberal cause. Yet while the PFP critically examined its leadership and electoral strategies, it rarely questioned the form its liberal democratic stance had assumed. The new grouping to the left which replaced the PFP, the Democratic Party, is still in the early stages of policy formulation but has clearly inherited the PFP's liberal democratic approach. Speaking of the new party, Zach de Beer, one of the triumvirate of leaders, declared: 'In two words, we believe in Western democracy'. He added: 'That democracy cannot succeed in diverse societies is a myth ... Clearly the Democratic Party has come just at the time democracy is most needed'.[1] In the popular consciousness, Western democracy is majority rule.

It seems an appropriate time to reconsider the constitutional approaches of the liberal opposition groupings: are their principles based on sound premises and are they viable in the divided society of South Africa?

Our examination does not relate to the liberal paradigm in general. The principle of optimum freedom and protection of the individual has, in our view, a timeless and universal validity. We direct our attention at the position taken by the liberal opposition amid the pressures and counter-pressures of contemporary South African society.

The common society reconsidered

In the years of classic apartheid, liberals rallied around two concepts, that of the common society and that of non-racialism. Initially the term 'common society' was more prevalent. In her presidential address to the South African Institute of Race Relations in 1956, Dr Ellen Hellmann gave her definition of the common society as one in which, 'by the systematic extension of the economic, social and political opportunities, all the rights society has to offer can be won by its citizens on the basis of achievement and worth but not on colour alone'.[2]

This definition did not rule out the concept of racial groups forming the building blocks of a political system. Nor did it suggest that a liberal democratic alternative to apartheid had to be instituted all at once. What was crucial was the idea of extending the area of non-discrimination. During this period the Progressive Party (PP) propagated a qualified franchise on a common roll, which would phase in the common society in stages. By setting educational and income qualifications it qualified its preference for individuals rather than groups.

By the early 1970s liberals perceived the need to develop a new liberal alternative to the apartheid system. All over the world the qualifed franchise had been discredited and had made its proponents vulnerable to accusations of élitism; even the homelands and other apartheid institutions in South Africa were based upon universal franchise.

The political commission of the Study Project on Christianity in Apartheid Society (Spro-cas) was set the task of developing a liberal policy which would provide a viable alternative to apartheid.[3] This commission, which was comprised of the best liberal talents in South Africa, stressed a staged approach, based on groups rather than individuals. Its report was accepted with near unanimity by the fourteen-member political commission.

According to its proposals, the first stage should be the extension of rights to members of all the black groups, by using or adapting existing institutions. It recommended:

- The development of the 'homeland' governments into viable forms of regional government which would not operate on a strict ethnic basis.
- The introduction of fully elected bodies of local government for the coloured people, Indians and Africans.
- The establishment of community councils which would serve as representative councils for coloured people, Indians and Africans, and which would also have legislative and executive powers for matters mainly affecting the communal group, such as education, health and welfare.
- The establishment of a multi-racial consultative body consisting of representatives of all groups. Its specific task would be to discuss basic political and constitutional issues in order to deter-

mine areas of consensus, possibly leading up to a national convention where a comprehensive new political system could be decided.

For the second stage the commission recommended the establishment of federal institutions, including a multi-racial federal assembly which was to be elected on an indirect basis (the various regional and communal authorities acting as electoral colleges) or by a combination of indirect and direct election.

In retrospect, it would appear as if the NP has been travelling along the road suggested by Spro-cas in some respects. However, Spro-cas never intended using official race classification (to which the government still clings) as anything but a starting point and a short-term expedient.

The Progressive Party did not adopt the Spro-cas proposals as its policy. For reasons which were perceived to be compelling they believed that the main task at hand was to propose a political system acceptable to the black majority, while at the same time securing some guarantees for the minority. Liberals had become apologetic; they had moved into retreat. But if acceptability to blacks was the main criterion then the message from the African National Congress (ANC), like that from nationalist movements north of the Limpopo, was quite clear: *if Africans had a decisive say*, they would totally reject the idea of any form of minority rights for whites or any other group.[4]

Furthermore the very idea of groups smacked too much of apartheid. By the mid-1970s, apartheid had lost any chance it ever had of being considered in Western eyes to be a competitive alternative to the non-racial model (see p. 100). Although a process of reform had been initiated, apartheid was condemned in ever more strident terms as being unequivocally evil. Opponents of apartheid declared that the conflict in South Africa was the result of white privilege and greed, and was not comparable with other nationalist or communal conflicts in which each side could rely on some understanding of its motives by the other side.

The new line of thought was broadly congruent with that of Edgar Brookes who, as a member of the Spro-cas political commission, strongly opposed its final report. While acknowledging the liberal thrust of the report, he was deeply troubled by its emphasis on groups. He wrote: 'But these are still groups, and groups based mainly on race or colour'. He concluded: 'The Commission, in short, proposes to cast out Beelzebub by Beelzebub'.[5] Within a remarkably short period of five years the minority view of Brookes became generally accepted by liberals and in particular by their new vanguard, the Progressive Federal Party (PFP).

The PFP adopted a new policy in 1978 after studying the recommendations of a constitutional committee headed by F. van Zyl Slabbert. In 1979 Slabbert became leader of the party and, shortly afterwards, published an academic defence of the policy in a book entitled *South Africa's Options*, co-authored with David Welsh.

The new policy abandoned the idea of group representation, as well as a staged approach to the introduction of non-racialism. It envisaged a number of federal states, each of which would be represented in the Upper House of a federation. The Lower House would be chosen on the basis of general adult franchise on a common roll within a system of proportional representation.

Instead of simple majoritarianism, the policy proposed a minority veto in which interested parties with more than 10 per cent representation would have the right to veto all measures except financial and administration matters and the election of the Prime Minister. As a concession to the group idea, it provided for the establishment of cultural councils to guard over the interests of cultural groups. No law affecting a cultural group could be passed unless it was first approved by its cultural council. The keynote aspect, however, was the majority-dominated Lower House — in the South African case the deeply symbolic House of Assembly, the fundamental seat of legislative power.

Drawing on local and international evidence, the Spro-cas commission had reluctantly decided against proposing a rapid transition to an open society on the basis of individual rights and representation. A mere six years later, Slabbert and Welsh (who had also drawn on academic studies of divided societies) concluded that South Africa's constitutional dilemma could only be resolved on the basis of individual rights and individual representation.

How did this shift come about? Since the hard truths about divided societies had not changed, the answer had to be sought in South Africa itself. What happened was that the Soweto uprising of 1976 had impressed upon South African liberal academics the need for a prompt accommodation with the black majority. The basic political requirements had appeared to shift. However in this they were quite at odds with the white electorate at large.

It is necessary to give somewhat deeper consideration to the propositions put forward by Slabbert and Welsh in *South Africa's Options* (and by the PFP) in order to assess its viability as a short- to medium-term strategy.

The assumptions of liberal democratic reformers

In the view of Slabbert and Welsh, the core of the South African conflict lies in inequality rather than a struggle between two competing nationalisms for security and identity. They argue that it is inequality which gives rise to the basic division in society between a power-holding group enjoying racial privileges and a subordinate group suffering racial discrimination. The authors chose not to allow for a specifically race-based minority veto on the grounds that this would simply perpetuate the status quo. In the words of the authors: 'The minority veto that we envisage would operate in the context of a constitutional dispensation that outlaws racial discrimination. It could not, therefore, be used to entrench something which is, by definition, illegal'.[6]

How could the support of the white electorate be won for a liberal democratic solution? The authors had three hopes. Firstly, they hoped that self-interest would lead parties or groups 'to negotiate with, rather than confront, one another'.[7] Secondly, they hoped that ethnic dimensions in the conflict could be regulated and controlled, and that many of the crucial issues could be depoliticized.[8] Finally, and most importantly, they put their hopes on the emergence of far-sighted leaders who would transcend parochial loyalties and would be able to negotiate a settlement that would lead the country away from racial domination to a liberal democracy.

These assumptions formed the basis of the liberal political paradigm through much of the 1980s. It was assumed that an increasing number of middle-class whites would come to accept that their property and other survival interests were ill-served by an unrepresentative white government. It was thought that once this happened whites would, to a growing extent, be prepared to rely on either a minority veto or the good sense of a black majority government to resolve the conflict of interests between whites and blacks in an equitable manner. It is important to note that while proponents of the liberal view are concerned with the protection of minority rights, they make no provision for mobilized minority *participation*. Protection without participation fails to address the identity needs of minorities in a political context.

Recent analyses of ethnic conflicts elsewhere, done by scholars such as Connor, Rothschild, Glazer and Horowitz, show that there is little evidence to lead one to believe that these assumptions are correct. The main insight of recent comparative studies of divided societies is that conflicts in these societies are not only about a clash of interests which can somehow be mediated. These analysts point to the dual nature of ethnic conflicts. On the one hand there is an interest component which, in divided societies, refers to the unequal access which the various ethnic groups have to resources. On the other hand, there is an emotional or identity component which refers to the need for communal or national identity, as expressed by political self-determination. Rothschild outlines it well: 'In sum, the relationship between the emotional component and the interest component of politicized ethnicity is dialectical. Neither is a mere epiphenomenon of the other.... and neither functions alone'.[9]

These scholars issue a strong warning against exaggerating the influence of the materialist component in politics. Connor has written that there is now abundant evidence world-wide that 'economic factors are likely to come a poor second when competing with the emotionalism of ethnic nationalism'.[10] And Cairns has pointed out in a study of the Northern Irish conflict, that the passions evoked by ethnic conflict far exceed what might be expected to flow from any fair reckoning of 'conflict of interests'.[11] The same applies to South Africa where severe repression and riot control operations stand in stark contrast to the fact that most whites in the middle and higher income brackets would probably not suffer a marked decline in living standards under a black government.

The new literature also rejects the tendency of a previous generation of scholars to regard ethnicity as being infinitely malleable in the hands of politicians. Horowitz points to the intense passion that ethnic allegiances regularly elicit, and warns: 'The ethnic group is not just a trade union'.[12] In similar terms Rothschild cautions that ethnic identities and group boundaries 'are not changed like shifts of underwear'.[13]

There is no reason to believe that it would be different in the case of white South Africans. Studies of Afrikaner attitudes in the mid-1970s showed that for the majority their concern with identity and concern with privilege were closely interwoven.[14] This sets distinct limits on the degree to which the ethnic identity of Afrikaners can be manipulated by a reformist leadership.

Even in the case of many white English-speakers one could not be certain of successful manipulation of ethnic boundaries.[15] Although the English-speakers are generally assumed to be materialistically motivated, one has to consider that they are also part of a white political community which is faced with pressures, sanctions and ostracism. In the 1987 election, in which national security was the main issue, almost half of the English vote went to the NP as the party of 'white nationalism'.

The NP's success rests on its ability to mesh the identity and materialist dimensions. The party's more articulate spokesmen express their constituency's concerns in the following terms: a combination of life-style, a sense of origin and identity, the psychological satisfaction of in-group community life, standards of public order, behaviour and responsibility, sufficient control over the allocation of resources, and the maintenance of security to secure the continuation of these benefits.[16]

Hence comparative evidence and theoretical perspectives raise serious doubts about the assumptions that materialist concerns would be decisive or that ethnicity could be manipulated and controlled by political leaders. We also need to address a third assumption of liberals, namely that competent leaders will emerge who will be able to break the log-jam in the event of ethnic conflict.

When violence engulfs a divided society there is an almost irresistible tendency, among both analysts and the public at large, to blame the leaders for the political malaise. In a 1976 poll taken in Northern Ireland, respondents were asked if they thought politicians helped or hindered the peace process; only 9 per cent said that they helped, while 63 per cent thought that they actually hindered the process.[17] Similar sentiments would probably be expressed in South Africa or Israel. There is a reason to wonder whether this is not a classic case of the need to find a scapegoat. In fact leaders with a popular power base are probably allowed little leeway by their electorates. Scholars in the field of conflict resolution have recently sought to put the concept of power in a more realistic perspective. Particular attention has been paid to the work of Paul Sites who attributed effective power not to governments, but to individuals and groups of individuals. He concluded that individuals would use all means at their disposal to pursue their own security and identity needs, subject only to the constraints they imposed upon themselves in their need to maintain valued relationships — that is, with members of their own ethnic group or political community.[18]

Given this perspective, it becomes obvious that it is unrealistic or even self-deluding to expect leaders to propose drastic policy changes and shifts in strategies to their followers. For instance, it is generally accepted that the ANC leadership would not be able to persuade its following to enter into structures unilaterally designed by the apartheid state. By the same token, no leader of a majority white party in South Africa would be able to win support for liberal democracy and its principle of one person one vote.

What this policy asks of whites is that they dissolve their successfully mobilized political grouping and enter into a competitive system with the black majority, in the hope that a party will emerge from elections that can attract sufficient support to block legislation. They are also asked to have faith that the black majority in power will not usurp the constitution and override the veto.

From the Afrikaners, in particular, it demands that they give up their close political identification with the South African state and its symbols,[19] and accept the rapid Africanization of the state. It is therefore not surprising that Afrikaner support for the PFP never rose to above 5 per cent. However exceptional a leader's qualifications may be, it is doubtful that he could persuade an ethnic group to abandon its political identity and to risk its material interests unless it perceived itself to be weak and vulnerable or without strategic options.

At the end of their book, Slabbert and Welsh concede that politics in South Africa will probably end up being the art of the impossible. While attaining and maintaining a democratic form of government in South Africa would be an enormously difficult task, they argue that it has to be attempted 'against all the odds'. [20] A critic has pointed to this basic flaw in the book: while the authors' own analysis of ethnic conflicts elsewhere and existing surveys give no real hope for a resolution based on a liberal democratic model of politics, the authors have resolutely stuck to their commitment to this model.[21] Had the authors used a different approach, and had they, for instance, investigated the historic preconditions for democracy, they may have come up with quite different policy prescriptions.

Preconditions for liberal democracy include the following: high rates of urbanization and industrialization, an advanced development of voluntary organizations, overlapping and cross-cutting social affiliations, widespread literacy, adequate and equitably distributed personal incomes and, above all, a widely shared sense of national identity. A noted Africanist has concluded that without such preconditions, liberal democracy was unlikely to work, and that Africa was, for the most part, without them.[22] A liberal democracy does not depend on good intentions but on social complexity and socio-economic development in a society.

At this stage South Africa lacks all, or virtually all, these preconditions. For instance, whites have control over 87 per cent of the land, more than 90 per cent of the total instruments of production and receive approximately 60 per cent of the total personal income. To immediately introduce multi-party electoral competition in order to crush or weaken one or another set of interests and needs, would merely exacerbate the existing racial and ethnic cleavages to the point at which democracy itself would be put at risk.

In academic theory, a politicized ethnicity can perhaps be avoided; in practice it runs counter to powerful, if mythical, needs and sentiments which are beyond logical persuasion. Furthermore, quite aside from these emotional factors, the sheer scope of inequality and the resultant clash of aspirations and interests, make authoritarian and anti-democratic political management virtually inevitable.

In Africa, a liberal democratic form of government has invariably given rise to open ethnic competition. It has generally weakened democratic processes and it has put the state at risk in many countries.[23]

We are not suggesting that these factors negate the ideal of democracy for all time. A fully and equally representative democracy must remain the goal, but its eventual emergence may be best served by an intermediate phase or a political transition which does not expose the system to an unmitigated clash of interests and needs.

A nation of minorities

In sharp contrast to the ideal of the common society, the National Party holds the view that whites are not a minority within a majority comprising mainly African people, but a minority alongside other ethnic minorities which are constituted on a tribal and linguistic basis. The notion of 'separate freedoms' for the multiple minorities, cherished by the National Party ideologues in the sixties, was grafted onto the older dispensation of 'reserves' for Africans which became the so-called homelands for the ethnic minorities.

This framework was applied not only to the then dominantly rural reserves but had its echoes in the administration of urban Africans as well. Multi-ethnic townships such as Soweto were developed according to the principle of separate neighbourhoods for the different 'tribes'. The policy of urban ethnic separation was never convincing and obviously was contradicted by inter-tribal marriage and non-tribal community organization. Today the old policy is reflected not in ethnic segregation but merely in different linguistic distributions in certain neighbourhoods in various townships.

The political implications of ethnic diversity among the African majority are no less complex and problematic than the issue of the common society. The two topics are more or less equally subject to wishful thinking and political artifice.

Ethnic identities do exist among the African majority. Seen very broadly, there is evidence that such identities have social and political consequences. It is true to say that under conditions of social and economic stress, linguistic and ethnic identity among Africans does become a vehicle for organized social action and a focus of conflict. The South African mining industry, for example, has a long history of sporadic but bloody conflicts between different linguistic/regional groups housed in the mine hostels.[24] Similarly, in Durban in 1985/6, there were particularly violent clashes between Zulus and Pondo migrants in certain peri-urban areas which left over 100 people dead.[25]

There is considerable disagreement among analysts as to the correct interpretation of these so-called tribal faction fights. The most accepted view among academic observers is that the root causes are not centred on ethnicity as such, but rather on work-place conditions, residential circumstances or competition for resources such as residential land. The fact remains, however, that it is remarkable how readily the emotions involved crystallize around ethnic identity in times of stress. This phenomenon is also well-known elsewhere in Africa.

There are more subtle factors associated with ethnic identity as well. Numerous empirical studies have found an interesting correlation between tribal group and political attitudes. For example: a 1987 study among 1004 African coalminers of largely rural origin in South Africa, conducted by the Emnid Institute of West Germany[26] found the following variations in support for the principle of majority rule. This was one of three options presented to respondents: it was contrasted with two arrangements for multi-racial power-sharing.

TABLE 5.3: SUPPORT FOR MAJORITY RULE OVER ETHNIC POWER-SHARING IN SOUTH AFRICA: RESULTS FOR SELECTED GROUPS.

Sample average	Zulu	North Sotho	Xhosa
27%	18%	24%	39%
(n 1004)	(n 272)	(n 129)	(n 120)

Source: Basic results of a study conducted for the Emnid Institute and the Deutsche Afrika Stiftung by IMS, SA, Ltd. Stratified random sample of black coalminers, July 1987.

Even in the urban industrial melting-pot, attitudes differ between ethnic categories. For example, in a 1986 sample-survey conducted by co-author Schlemmer among 1333 African workers in seventeen large industrial concerns, ethnic differences emerged with regard to political

TABLE 5.4: PROPORTIONS ENDORSING THE ANSWER CATEGORY OF 'ANGRY AND IMPATIENT' WITH REGARD TO FEELINGS ABOUT THE POLITICAL SITUATION IN SOUTH AFRICA.

Sample average	Zulu	Xhosa	Sotho	Other
45%	62%	44%	45%	30%
(n 1333)	(n 392)	(n 357)	(n 421)	(n 163)

Source: Basic results of a study conducted for a major corporation. Fieldwork by IMS, SA, Ltd. Stratified random sample of black industrial workers, September 1986.

consciousness (Table 5.4).

The variations above could not be explained on the basis of educational level, level of urbanization, age or any other basic variable.[27]

The relevance of linguistic identity to politics has also been assessed more directly in opinion survey questions. The following findings are excerpts from a review of evidence,[28] based on a question posed to multi-ethnic black samples in 1979 and 1981. People were asked: 'If all black people (i.e. African) voted in South Africa, (name of group) would be small in number and other groups might have more power in government. Would people like you feel weak and insecure or do you agree with the statement "blacks are all one people — it would not matter at all"?'

TABLE 5.5: THE 'WEAK AND INSECURE' OR DEFENSIVE ETHNIC POSITION WAS ENDORSED AS FOLLOWS:

Soweto Xhosas	24%	(n 100)
Soweto Zulu	21%	(n 75)
Soweto Pedi	35%	(n 75)
Rural Xhosa	44%	(n 100)
Eastern Cape Xhosa	50%	(n 417)
Rural Zulu	47%	(n 250)
Zulu migrant workers	64%	(n 100)

Source: L. Schlemmer, 'Data collection among black people in South Africa', *HSRC bulletin*, vol. 13, no. 1, 1983.

These results suggest that an association between ethnicity and politics is present, but that it is more salient in ethnically homogeneous regions such as the Eastern Cape or in rural areas than it is in the urban melting-pot. One obviously cannot generalize from such relatively flimsy evidence but these results might suggest that in a future majority rule dispensation, unless strong counter-mobilization occurs, there could be a tendency for regionally-based or ethnic dissent to develop, such as that which has caused conflict in Zimbabwe, Angola and elsewhere in Africa.

In research conducted for the Buthelezi Commission in the early 1980s by co-author Schlemmer, an attempt was made to probe general principles relating to ethnicity in politics by posing questions in regard to other African countries. There was considerable support for the recognition of African ethnic minorities in government and for the protection of language diversity. Even among samples in multi-ethnic Johannesburg this sentiment was evident in a majority in all ethnic sub-samples. The conclusion drawn from this is that 'there is substantial minority pride in ethnic identity, rather less aggressive tribalism but near majority concern that ethnic characteristics should not be obliterated in a political dispensation'.[29]

These results notwithstanding, in a situation in which all ethnic categories among Africans feel more or less united in protest against the common factors of their exclusion from Parliament, it is easy to find evidence supporting inter-ethnic political unity at the grassroots level. From the evidence from the Buthelezi Commission reviewed earlier,[30] one response did emerge as typical: 'There is no real difference between African people — we can easily forget tribal differences and stand together. Do you agree or disagree?' The percentage *agreement* among all ethnic and regional categories in a sample of 750, exceeded 80 per cent.

What all these results suggest is that ethnicity may well have political effects, perhaps even dramatic effects, in a future unitary state (unless it is dealt with sensitively), but that African ethnic identities are not absolute social or invariant 'primordial' features. The salience of ethnicity varies widely depending on the region and the extent to which it has become incorporated into other aspects of social or occupational life. Most importantly, at present there is a *duality* of consciousness concerning ethnic identity. In other words, ethnic concerns are swept aside in favour of African unity in the context of a common struggle for rights, particularly among unenfranchised people outside of the homelands.

The policy of separate development (which was aimed at political autonomy for each African ethnic group) assumed a rigid, more or less automatic, ethnic self-classification and was manifestly over-prescriptive to say the least. It was a type of imposed ethnic engineering that more than fully deserved the odium which it earned.

We need say no more than this, if for no other reason than the fact that government policy has shifted away from the ethnic categorization of Africans. The homelands are taken as a *fait accompli* and are now, to all intents and purposes, peripheral dependencies of South Africa, whatever their ethnic attributes may be.

As regards Africans outside of the homelands, the government has cast aside ethnic considerations in favour of planning regions and local authority areas as a basis for the political incorporation of Africans. In the National Council proposed by government there will be nine African representatives for the common area, probably one from each of the economic development regions, elected by African local authorities in each region.[31] Since some of these regions are multi-ethnic, the tribal identity of possible electees will be fortuitous.

Hence the issue of African ethnic minority rights can no longer be seen as highly salient political issues for government or anybody else. The emerging issue is the debate about unitary black participation in a non-racial dispensation, as proposed by the ANC/ UDF and, for that matter, Inkatha and the Black Consciousness Movement, versus the system proposed by government.

The government's approach is asymmetrical since it envisages standard constituency-based representation of whites, coloureds and Indians (albeit separately) as opposed to regionally based representation of Africans. Groups of townships in a region, as well as homelands, are expected to nominate representatives to a national council. The regional basis is obviously a response to the reality of the homelands as well as the numerical preponderance of Africans. Regional representation is a device to neutralize demographic differentials. Whatever the case, however, the old principle of tribal self-determination has been eclipsed. Clearly neither liberal democracy nor the 'nation of minorities' theory provides a solution which can attract strategic support from both sides of the political divide.

We proceed now to a discussion of the major change strategies. None of them has a specific constitutional solution but each has advocates which consider it a suitable method by which South Africa can be shifted dramatically in the direction of a political system acceptable to the majority of South Africans. We review firstly the prospects for a revolution before turning to a discussion of sanctions and then to an internationally-backed settlement along the lines of the Lancaster House agreement or Resolution 435.

Revolution

A persistent view of South Africa in the consciousness of most opposition activists and overseas observers, is that an inevitable, massive insurrection will force the regime to capitulate or dramatically accelerate the pace of concessions. This perception has spawned a plethora of 'crisis' literature on South Africa as the following titles signify: *South Africa: Time Running out*,[32] *Towards a Certain Future*,[33] *South Africa at War*,[34] *Endgame in South Africa?*.[35]

These views have not been so evident since the latest cycle of rebellion in South Africa subsided under the impact of the draconian legislation imposed by the State of Emergency in 1986. Nevertheless this remains a powerful implicit paradigm within which the potential change process is viewed.

We argue that there is little convincing evidence to support the theory that South Africa will become sufficiently destabilized to force the government to negotiate the transfer of power. Between 1984 and 1986, the state survived the most dramatic cycle of protest and internal rebellion that the country has known. Although this revolt has been suppressed, many observers and activists still believe that the events of 1984 – 6 demonstrate the intrinsic vulnerability of the South African state and that its future overthrow could be achieved by a better organized combination of internal resistance and external pressure.

It is necessary to put the events of 1984 – 6 in a proper perspective. We would argue that if one were writing a textbook on how a government could invite a political upheaval, one would identify the following steps — all of which the government took between 1983 and 1986:

- initiate constitutional reforms just when the economy takes a sharp dip; reforms which exclude Africans and increase their sense of relative deprivation;
- ease up on security action in an attempt to forestall sanctions and to give the new constitution some legitimacy, regardless of the impression this gives of vulnerability to foreign pressure and protest, thereby allowing the mobilization of black youths who, as youths, are least inclined to political self-discipline;
- introduce municipal 'reforms' which fragment what little sound administration exists at local government level;
- force the newly elected black town councils to increase rents and service levies;
- when disturbances break out, reimpose restrictions on, and avoid negotiating with, the leaders and organizations which have a capacity to exercise discipline on the youth.

Need one say more! It would have been a miracle indeed if South Africa had not gone through a period of political violence. More importantly for our argument, however, is the fact that it is unlikely that a similar combination of mistakes will be repeated.

If one looks beyond the specific events of 1984 – 6, to a more abstract analysis of the conditions under which states are overthrown, it becomes clear that the South African state is an unlikely candidate for revolution. One of the most outstanding recent theoretical advances in the understanding of revolutions, can be found in the work of Theda Skocpol. Skocpol takes issue with scholars who believe that revolution is inevitable once a majority gets disgruntled and that states are incapable of repressing a discontented majority for long. Skocpol argues that these views are naïve. As she phrases it, 'They are belied in the most obvious fashion by the prolonged survival of such blatantly repressive and domestically illegitimate regimes as the South African'.[36]

Skocpol concludes that even after a great loss of legitimacy a state can remain quite stable — as long as the security forces remain coherent and effective in controlling the population, and the state continues to collect sufficient taxes and recruit enough policemen and soldiers. The state must also be autonomous in the international system, in the sense that it cannot be brought down by a colonial power (as was the case in Algeria) or by the withdrawal of support of a neighbouring patron (as South Africa did in Rhodesia) or successful economic sanctions.

How does the South African state rate on these scores? The state is experiencing no difficulty in finding sufficient black recruits for its security forces. White conscripts are reluctant but acquiescent. Furthermore, there is no evidence of a break-down in discipline or a refusal on the part of the security forces to take action against turbulent crowds. Studies show that there is little or no prospect of a revolution before a significant section of the security forces transfer their allegiance to the resistance movement.

The South African state does suffer considerable budgetary constraints. This is not surprising, given the demands of a white electorate accustomed to First World standards, as well as a rapidly growing black population with huge expectations of the state.

However, there is no indication that the South African state is unable to gather the required minimum revenue. Towards the end of the 1980s, the state considerably enhanced its capacity to gather revenue. Prior to the 1989 budget, the rabidly pro-business magazine, *Financial Mail*, wrote scathingly of an 'unending onslaught that has led to a flood of funds into Treasury coffers'. Borrowing was expected to be lower than the budgeted 4,9 per cent of GDP, mainly as a result of revenue running more than 25 per cent ahead of budget.[37] Even though the government has lowered personal income taxes in an attempt to encourage savings and entrepreneurship, indirect taxes and a spreading of the tax-burden, as recommended by the recent Margo Commission, will ensure an adequate flow of revenue.

There is no doubt about the state having the means to defend itself. It is generally recognized that the South African armed forces are vastly superior to the ANC's and PAC's military wings. The government, of its own choice, is not spending inordinate amounts on security. The defence force is neither particularly large nor particularly expensive in a world context. South Africa's military expenditure at 13,4 per cent of government expenditure (1983) is well below the United States (29,6 per cent), Taiwan (25,9 per cent) and Israel (24,9 per cent).

By any standard South Africa is still under-policed. In June 1984, a government spokesman indicated that there were 1,4 policemen per 1 000 of the population which compared with roughly 2,3 per 1 000 in the United Kingdom, 2,7 in Israel, 4,4 in Northern Ireland and 5,7 in Algeria. South Africa's police strength has recently risen to roughly 2,0 per 1 000. Nevertheless there are few countries with a lower figure for police per 1 000 members of the population.[38] In other words, there is scope for increases in the strength of security forces.

To be effective, security forces need to be able to fall back on a cohesive ruling class or ethnic group. Even a relatively weak popular opposition may overthrow a ruling group which is torn by conflicts, reluctant to use force or is lacking in political skills.

That the South African state recognizes this is clear from the words of ex-police commissioner, Johan Coetzee, who has characterized the conflict in South Africa as a war of attrition. He observed: 'He who lasts longest wins, because in the final analysis, it is about the will of the nation to exist and that of others to wear them down'.[39] As Harvard's Samuel Huntington remarked: 'Revolutionary violence does not have to be successful to be effective. It simply has to create sufficient trouble to cause divisions among the dominant group over ways to deal with it'.[40]

In 1985 and 1986 there appeared to be a break-down in the cohesion of the white ruling group. The primary base of government, the Afrikaners, had split politically and there were signs of disaffection among white business leaders, academics, students and churches. In mid-1986, Stanley Uys, a respected political commentator, anticipated a mighty upwelling of black rage that 'will sap and haemorrhage the white power structure until it falls in on itself'.[41]

But the signs of political disarray and confusion are deceptive. Behind disagreements over tactics and strategy, the whites in general, and Afrikaners in particular, harbour a formidable will to survive politically. A recent study of student attitudes at the University of Stellenbosch, the most open-minded of the Afrikaans campuses, found that, while 32 per cent of students supported parties to the left of the NP only a very small proportion could be considered fundamentally alienated from the state.

Four-fifths of the students had no problem in condoning repressive action by the police, such as the breaking up of peaceful demonstrations, the shooting of demonstrators who damaged property and the detention of people who participated in protest marches. About the same proportion was hostile to the ANC and UDF. Some 10 per cent showed sympathy but not more than latent support for the ANC and UDF. Active support for these organizations was down to 3 per cent and appeared to be dwindling. By contrast there was strong majority sympathy for homeland leaders, Inkatha and the incumbents of the Indian and coloured Houses of Parliament, which are movements and parties prepared to work within the system.[42]

The final necessary condition for revolution is black unity. Although virtually all blacks oppose apartheid there are deep-seated differences over political strategy. The tensions between 'Black Consciousness' and UDF-linked activists, the bitter and often bloody battles between Inkatha and the UDF, particularly in the Pietermaritzburg region, and the conflicts between different groups of squatters (for instance, in the Guguletu-Nyanga-Crossroads complex in Cape Town), and those between black policemen and rock-throwing youths, have dissipated black energies which could otherwise have been directed against the apartheid order. The state has been able to exploit the divisions or, as in the case of the Pietermaritzburg conflict, has allowed the fighting to continue, thereby weakening black opposition forces.

Some observers concede that a revolution is unlikely at this stage but argue that revolutionary potential is slowly and inexorably mounting, or that cycles of unrest or rebellion are following one another more rapidly.[43] These views often reflect the observer's own projection onto the situation or a naïvely mechanical view of social processes. Even if one assumes a steadily intensifying political awareness among black people, one also has to accept that all behaviour is substantially motivated by perceptions of the likelihood of success or reward, and that even an intense political consciousness has to be mobilized in order to be effective.

Repressive security action which is consistent and pervasive is likely to depress motivation and suppress mobilization. In the aftermath of the events of 1983 – 6 the internal black resistance movements have been weakened and fragmented. The security system appears to have become more sophisticated and more effectively co-ordinated with civilian bureaucracies through so-called Joint Management Centres (see Chapter 4).

We accept that there is a high degree of political consciousness, a latent protest and an ever-present potential for mobilization in South Africa's townships. Notwithstanding the success of the mass 'stayaway' strikes of June 1988, the de-escalation of sustained protests under the impact of the second State of Emergency bears out the points made about security action above. Thus there appears to be very little basis for assuming that the government will be forced to negotiate because of the impossibility of controlling the internal situation.

Popular resistance has suffered from what can be called the debilitating effects of an *illusion* of power. John Kenneth Galbraith warns: 'Nothing so serves the military or corporate power as the belief of its opponents that they had accomplished something by holding a meeting, giving a speech, or issuing a manifesto'.[44] To launch a serious challenge, the resistance movement will have to deploy its considerable resources in viable strategies which pursue realistic goals.

Sanctions

Among the various aims of the international sanctions campaign is the intention to force the South African government to negotiate by imposing costs on the economy. Leaving aside such controversial questions as whether or not a majority of blacks want sanctions, and whether or not the costs of sanctions can be transferred to the black poor and to the economically dependent Frontline States, we are left with probably the most important question: do sanctions have the necessary political impact to facilitate or force negotiation?

The term sanction implies pressure. Can sufficient pressure cause behaviour to be modified? In the South African case, will the costs of this be bearable?

In an interview with co-author Giliomee in October 1988, a main strategist of the ANC expressed the view that within six months of comprehensive mandatory sanctions, a Botha or his successor would be sitting down at the negotiating table with the ANC on a realistic agenda.

Comprehensive mandatory sanctions would undoubtedly constitute a major blow for the South African state. At the same time, however, all indications are that, among whites, such a step would generate a combination of fatalistic determination and nationalist fervour. One could expect a refusal to capitulate similar to the responses of civilian populations during the Second World War, during the Blitz in Britain or the saturation bombing of German cities.

No single careful analysis of which we are aware suggests that sanctions will cripple the South African economy in the short to medium term.[45] They are, however, imposing economic costs — the premiums paid to middle-men in evading sanctions, impediments to the flow of technology, scarcity or absence of external capital and trade credit, a generally lowered export performance and, because of balance of payments problems, downward pressure on the rand.

At the same time, however, certain steps taken by domestic economic decision-makers could in part offset the effects of sanctions in the short to medium term. Import replacement strategies could stimulate a diversification of domestic production, some of which, because of a low rand exchange rate, could be exported via the international black market. This is currently the case with the South African arms industry.

The high costs of imported equipment could encourage less capital-intensive production, hence stimulating job creation. Furthermore, as increasing numbers of overseas firms disinvest, their manufacturing plant, expertise and local capital will become available to South African purchasers at bargain prices.

What, then, will be the impact of sanctions? Although undoubtedly serious as regards longer term growth prospects — capital shortages will be a major factor in the long run — the pressure on the economy due to sanctions will not be catastrophic. A slow wearing down of both main contending forces within South Africa is the most probable result of international sanctions.

In view of the fact that the pressures will fall short of engendering an economic crisis in the short to medium term, what are the political reactions to sanctions likely to be? This is a crucial question, since the outcome as regards negotiations hinges upon the answer.

There is an equally important supplementary question: is the government critically dependent on a support-base whose interests are likely to be seriously harmed by sanctions? The government's supporters are by-and-large middle to lower-middle class whites in lower managerial, supervisory and clerical positions in the private and public sectors. They are not the very wealthy whites who have a taste for imported goods or the major industrialists dependent on export markets.

Those in the private sector are unlikely to lose their jobs through retrenchment since South Africa has an overall shortage of experienced middle managers and supervisors. Those in the civil service are harmed more by government attempts to reduce its spending in order to lower the inflation rate than by sanctions. If the state has to step into the economy to stimulate strategic production, many members of the civil service could increase their opportunities and sense of importance.

Most whites do not favour concessions to the sanctions campaign. In a poll conducted among a representative probability sample of 1808 whites in 1986, undertaken by co-author Schlemmer in association with Market and Opinion Surveys (Pty.) Ltd., the question was posed: 'How do you view the attempts in America and elsewhere to influence our policies — what should be our reaction?' The poll revealed that 38 per cent of the white general public considered that policy adjustments were inevitable, as against 35 per cent of government supporters, and under 2 per cent of the more conservative supporters of the Conservative Party (CP) and the Herstigte National Party (HNP). Only the supporters of the liberal Progressive Federal Party (PFP), which attracted less than 20 per cent of electoral support, produced a majority (69 per cent) which favoured policy adjustments in response to sanctions. The majority is likely to be smaller in the Democratic Party (DP) which has replaced the PFP, because the DP includes some fairly conservative former New Republic Party (NRP) voters.

If only one third of government supporters considered 'policy adjustments' appropriate, the proportion which would have gone so far as to support actual negotiations over power with the resistance movements would have been much lower than this. These results were obtained despite the fact that 70 per cent of government supporters feared that sanctions would undermine the economy, as opposed to only 3 per cent who felt that sanctions would not actually affect them.

The people who would be most critically affected by sanctions — black production workers in export industries — are not likely to exercise any leverage over government. They might do so through their trade unions, but the current trade union leadership is unlikely to engage the government in a debate over the type of compromise solutions which would avoid or minimize sanctions. In fact, trade union leaders display a political enthusiasm for sanctions, despite the fact that this is somewhat at odds with their members' economic interests. Obviously the trade unions will attempt to encourage the government to negotiate with the ANC, but this has been their stance for some time, with little effect. If their power-base is weakened through sanctions, resulting in growing numbers of black workers being unemployed, then encouragement of negotiation would count for even less.

Aside from these matters of economic interest, there are political dynamics inherent in the sanctions campaign which incline us to pessimism regarding its effects. Firstly, sanctions will tend to allow the government to blame external pressures for problems which might have little direct connection with sanctions. In particular, sanctions will give the government opportunities to shift the blame for economic problems onto external scapegoats.

Secondly, most whites feel that the sanctions campaign has intensified in recent years despite reforms such as the legal recognition of black trade unions, the abolition of passes and the recent government proposals to open certain selected areas to residents of mixed race. One gains the impression that whites increasingly see the campaign as containing non-negotiable demands — and ultimately a demand for complete capitulation. Attitudes of hostility to, and categorical rejection of, the campaign seem to be gaining ground among whites.

Thirdly, the internal proponents of sanctions, such as Archbishop Tutu or (more cautiously) certain leaders of the trade union movement, are not regarded by government as spokespeople for rank-and-file blacks on the issue. The perception is that these spokespeople are committed to political goals and strategies which ordinary black people do not endorse. The fact that most opinion polls (ten out of fifteen) show that most blacks reject sanctions is a factor in this process.[46]

In fact, a recent survey of 1004 black coalminers, conducted under the supervision of the West German research organization Emnid, showed that only some 28 per cent of members of the National Union of Mineworkers (NUM) supported sanctions (13 per cent among non-union members). NUM, the single most powerful component of COSATU, has endorsed sanctions several times, yet only some 40 per cent of black coalminers were aware of this stance. Thus any claim by NUM that it supports sanctions with the endorsement of its members lacks credibility.[47]

These factors and findings appear to indicate to both whites and members of government that sanctions are 'illegitimately' endorsed from within South Africa. The pro-sanctions view is seen as radical and unrepresentative. From comments by respondents in a survey of 1804 whites, conducted in November 1986 by Market and Opinion Surveys, it seemed that only some 15 per cent of whites felt that

Archbishop Tutu was acting legitimately in expressing the view he does on sanctions.

Fourthly, the most telling symbolic accusations of the right-wing official opposition party include the criticism that the government is attempting to appease external opinion. Given the fact that the right-wing opposition appears to have the support of over one-third of the white electorate, the government is likely to feel that it cannot afford to give the opposition too much additional ammunition by making concessions to external pressure.

Finally, if the sanctions campaign, by creating an impression among black South Africans that external allies are fighting their political battles for them, reduces the energy and determination with which black organizations inside the country put their case to government, then the campaign will indeed have done more harm than good. The external sanctions campaign may also artificially improve the morale and confidence of black movements, and they may become less amenable to compromise. The case is impossible to prove either way, but this is a worrying aspect of the sanctions campaign.

Among the benefits of the overseas sanctions campaign is that it probably helps to focus the attention of both whites and the government on the fact that South Africa's policy of socio-political differentiation on the formal basis of race flaunts universal political morality. Sanctions do maintain the moral pressure for change and, at a time when internal protests from blacks are subdued, prevent a perception taking root again among whites that their policies are normal and appropriate in an ethnically plural society. On balance, however, there is almost nothing to suggest that sanctions will help to bring about serious negotiations over the political rights of the majority in the foreseeable future.

There is perhaps a way in which external pressures could facilitate such negotiations. If the present level of sanctions were regarded as a platform on which to base more creative pressures, results could be achieved. For example, if Western governments were to link a series of realistic reforms and steps towards negotiations with a selective lifting of sanctions and with specific forms of economic relief, such as unfreezing external credit facilities, effective pressure on the government from inside South Africa would increase. A sense of obligation to respond to 'reasonable' expectations by the West could be created.

At present, however, sanctions have an image among most whites of unmitigated and fairly boundless hostility. As such they create a solidarity in opposition to them. The very heat of the sanctions campaign and the unreflective fervour of its internal and external exponents mean that opportunities for creating effective pressures are lost. The goal of a negotiated settlement does not appear to be well-served by the sanctions campaign.

An internationally-backed settlement

In an article published in early 1989, R. W. Johnson, the author of the classic study *How Long Will South Africa Survive?* (1977) wrote that the day may well come when the South African government tries to extricate itself from a 'messy, uncontrolled drift towards an increasingly alarming future'. Johnson continues: 'A rational path for the State President would be to approach the G7 [Group of Seven, i.e. the major industrialized nations] as a bloc and work out with them a phased programme of transition to majority rule, the resulting settlement to include provision for the protection of minority and individual civil rights, to be guaranteed for a period of, say, ten years by the G7 powers'.[48]

Recently, thoughts similar to those expressed by Johnson have exercised the minds of people in power in London, Bonn and Washington. The basic question which they address is: if neither internal nor external pressures can force Pretoria to negotiate a transfer of power, would both types of pressure plus high-key and influential external or international facilitation produce the necessary incentives for Pretoria to accept a political deal?

One fairly similar instance in which this combination of pressures was successful was that of the old Rhodesia in the immediate pre-independence period. The Rhodesian settlement is frequently taken as a model of what could occur in South Africa and has inspired a range of initiatives and strategic policies.

We shall digress somewhat to discuss the Rhodesian situation as a possible analogy. Before 1978 the prospects for settlement in the old Rhodesia appeared to be virtually as unpromising as in present-day South Africa. Nevertheless a series of events occurred shortly afterwards which did yield a resolution surprisingly quickly and it may be instructive to consider that process for the lessons it may contain.[49]

The Pearce Commission in Rhodesia had revealed the deepest polarization between contenders to power, while the heightened offensive by the Rhodesian security forces from 1977 onwards had given whites a confident feeling that they could push back the guerrillas. At the same time the South African government, for various internal political reasons, had ceased to apply manifest pressures for settlement. Nevertheless, certain external initiatives and internal changes finally resulted in altered perceptions on both sides which finally made the settlement in 1979 possible. In the mix of factors, both before and during the Lancaster House process, the following seem to us to have been particularly significant:

- The tough, unsentimental, honest broker role played by Britain under Lord Carrington proved to be more credible than the earlier initiatives by Kissinger and Carter.
- Very importantly, interim changes and shifts had occurred which fragmented the earlier sharp polarization. In terms of the Kissinger package, the white government in 1976 had agreed in principle to majority rule in ten years. Internal power-sharing under Bishop Muzorewa, although not a solution in itself, changed perceptions of black participation among whites.
- There were strong pressures from the Frontline States on both Mugabe and Nkomo to modify their fairly uncompromising stances on negotiation. In particular, the economic damage done to Mozambique by the war was significant in this regard.
- Furthermore, indications that Britain might recognize Muzorewa as leader of a viable government, moved the leaders of the guerrilla movements to consider compromise as a way of heading off this possibility.
- Bishop Muzorewa and his lieutenants were also important in that the British could use them as a channel for appealing to Smith to display more flexibility. By the time of Lancaster House, Muzorewa and his advisers had become an important element in the political establishment and their willingness to work with the British was vitally important.
- The British were also able to appeal to the Rhodesian military chiefs with some success, since these generals knew that the war was a hopeless cause in the long run.
- During the Lancaster House negotiations there was an ever-present threat of withdrawal by the Patriotic Front (PF) leadership, who felt that they were being dragooned into a settlement which fell short of various important demands. This threat was headed off by the frontline states, with which Britain maintained intensive communication. The Frontline States, as base territories for the guerrillas, had considerable leverage with the PF leadership.
- The fact that the Lancaster House proposals made provision for a British governor to take over executive authority in a transition

leading up to the elections was immensely reassuring for whites, who would have balked at the PF entering or taking over government directly.

- Furthermore, for the whites the presence of South Africa was also vital as it offered an escape route and the assurance of assistance should the cease-fire break down.

These points, as Robert Jaster comments, indicate that the dispute had to evolve through a number of stages before a settlement was possible. The attainment of greater symmetry in the negotiating positions and leverage of the contending parties was one reason why the stages were necessary. It was also necessary for the basic demands on both sides of the conflict to be taken seriously, although this only emerged after a number of failed initiatives. The great importance of constructive pressures from neighbouring states was a further element. Finally, no settlement would have been possible without the British who were considered to be impartial mediators and who had considerable leverage on both sides.

Clearly, there are also very large differences between the present situation in South Africa and the last years of Rhodesia. Among the most significant is the fact that there is no impartial external mediating power with anywhere near the same relevance to the situation as Britain was to the Rhodesian conflict. Secondly, the guerrilla war, that proved to be unwinnable for Rhodesia, has not even started in South Africa.

At this stage the armed conflict in South Africa consists of sporadic incursions, not war, and the defence force is still well able (through its destablization tactics) to effectively discourage Frontline States from accommodating full-scale guerrilla encampments. Furthermore, South Africa does not have what the white Rhodesians had: South Africa as a final guarantor of safety and ultimate refuge. Encouragement from South Africa on Rhodesia to settle was also very significant in the final stages of the negotiations surrounding the Rhodesian conflict.

This brief comparison makes it clear that the conflict in South Africa has to evolve through many more phases before the time is ripe for a Rhodesian-type of coerced settlement, or for power-plays similar to those which occurred at the end of the rule of the Shah in Iran. The breakdown of the comparison also raises the question of whether or not there is any point in focusing on a 'settlement' when negotiations over power seem so distant.

Namibia and the progress currently being made with a settlement in terms of United Nations Resolution 435 is also suggested as a precursor of a similar 'settlement' in South Africa. Although a fascinating topic in its own right, for present purposes the topic need not detain us for very long.

Namibia's tiny white minority, fifty times smaller than the South African white population, has been even more of an appendage of the South African state than was the case with the old Rhodesia. The Namibian whites had no more than a fraction of the resources needed to set their own political agenda compared to South Africa's white minority. Hence when President P. W. Botha did a cost-benefit analysis and decided that Namibia was dispensable, the issue was settled. South Africa was prepared to embark on the implementation of Resolution 435 for several reasons.

There were not only mounting military costs, but the war in Angola (and the UNITA movement) had helped to ensure the effective neutrality of an independent Namibia. To this must be added the strategic gains Pretoria stood to make, firstly in the agreement that ANC bases would be removed from Angola, and secondly, from the support it got for independence from the Democratic Turnhalle Alliance (DTA). If the DTA features predominantly in the future Namibian politics, Pretoria will have an ally in that territory. The situation has few parallels with the internal situation in South Africa itself.

Thus, after our brief review of three major popular models of transformation for South Africa (the revolutionary model, the external pressure equation and the settlement model drawn from the examples of neighbouring states) we have to conclude that the major dynamic for further change, and the one with which external pressures will have to interact, is internal pressure. For this reason we turn now to the question of how change will actually occur in South Africa. It is our contention that unless the major opposition forces, both inside and outside South Africa, understand this, their efforts will be in vain or even counter-productive. If the international community wishes 'to do something about apartheid', it needs to start with an in-depth analysis of the process of change in South Africa.

Effective pressure for change

In Chapter 4 we broadly sketched the changes which have already occurred. We have addressed issues, interests, pressures and outcomes which hold the key to what is likely to happen in the future. We now shift gear and assess the intervening processes which may well carry South Africa beyond the untidy adaptation which we have termed reform-apartheid.

Steven Friedman identifies a range of implicit theories of change which inform the current debate about strategies in South Africa. These are what he terms the 'white change of heart', foreign intervention, violent overthrow of the present order and internal resistance which is predominently non-violent.[50]

Friedman believes that internal (non-violent) resistance is the dynamic which has been most effective in shifting government policy. Appropriately, he qualifies the process, arguing that it is not necessarily effective on its own, but that it is more salient when organized and when the organizations involved have entered into tactical alliances with influential white business groups.

He singles out the successful break-through of the black labour movement as one good example. He also mentions black squatter resistance as a significant factor in the government's retreat from the pass laws. He argues that the 1976 protests were also a factor which helped to move government to a recognition of the permanency of blacks in the common area of the society. He is cautious, however, and warns that although well-organized and strategic black resistance can and has forced concessions it has not threatened the power of government.

On the other hand, he is optimistic in his view that the erosion of white control has already begun and that further erosion is a real possibility. For him, the effectiveness of black organizations is the crucial factor.

To this assessment we can add Friedman's earlier analysis which suggested that the process of change acts from marginal issues inwards towards the core issues of power. As concessions are made regarding peripheral issues — and made more readily because of the marginality of the problem — issues closer to the core of power become vulnerable to the effects of pressure. Hence, change is a strategic sequence, moving from softer issues ever closer to the core interests of the white ruling group.[51]

In recent years, opportunities have opened for a another type of pressure for change which could be called the 'mobilization of information'. Under the leadership of P. W. Botha, a 'technocratic' mode of problem-solving and policy formulation emerged.[52] Increasing reliance came to be placed on information-based strategies by the professional, civilian and security bureaucracies. A senior official in the government Bureau for Information recently told co-author Schlemmer that the cabinet and certain of its committees sometimes wait for the results of opinion-surveys before taking decisions.

This new mode of operation is likely to persist under a new president, simply because it offers great benefits compared to other ways of dealing with a complex crisis situation. Amongst other factors, the mobilization of information by the Urban Foundation, by scholars,[53] and a number of business-linked organizations, as well as by the President's Council itself in its signal report on urbanization, contributed to the very important step of dismantling influx control regulations in 1986.

A further important dynamic of change was established by the government's own constitutional reforms. It is now clear that the new constitutional initiatives have had unforseen consequences: for example, the introduction of the Tricameral Parliament stimulated reactions which led to the formation of the UDF, thus giving the exiled ANC a partner organization supporting the Freedom Charter inside the country. The government's constitutional adjustments have also opened up other sources of leverage for change.

Any government will attempt to adapt policies in order to protect the institutions it creates. The state has mobilized socio-economic reform in order to develop the legitimacy of the black local authority system, as part of its drive for stabilization. It is manifestly clear that the African, coloured and Indian political figures who operate inside the political system on the level of local, regional and national government, have considerable influence in their roles as custodians of the new structures. If they do not have direct leverage, they certainly have substantial scope for spoiling tactics.

This is perhaps the most notable departure from the original Verwoerdian apartheid model. Because it was intended, at that stage, that all homelands become as separate as possible, the central state machine and the constitutional system were completely insulated. In the current phase the non-independent homelands are essentially part of the emerging 'multi-racial' constitutional arrangement — in fact they are part of the mechanism for extending it into the black constituencies.

In other words, in recent years, strategic or qualified participation in state structures has begun to emerge as a rather potent source of legitimate pressure on government which, by definition, it cannot ignore or undermine. An example of this is the pressure presently being exerted by Alan Hendrickse of the Labour Party who insists on sweeping changes to the Group Areas Act.

In a similar context, Chief Mangosutho Buthelezi's refusal to participate in the establishment of Regional Services Councils in parts of Natal constitutes a problem for government, which it might only be able to resolve through further political concessions. For instance, the KwaZulu-Natal Indaba, promoted by KwaZulu and established jointly by KwaZulu and the Natal Provincial Council (in its earlier elected form) has tabled constitutional proposals for a unified form of regional government. Furthermore, negotiations on an adapted form of Regional Services Council are already underway.

The Indaba is persisting with its proposals and government will have to accommodate them in some form sooner or later. Although government has not responded formally to the Indaba proposals, a joint committee was established between KwaZulu and the Department of Constitutional Development, in February 1989, to explore prospects for and impediments to, negotiations between the central government and the state government of KwaZulu. The government is expected to present its formal reaction to the Indaba proposals fairly soon, after which discussions may follow. Mr F. W. de Klerk and chief Buthelezi have already met.

At a somewhat different level, it has been observed that white South Africans, including the constituencies supportive of government, have well-developed and understandable anxieties about the future. This makes them particularly receptive to ideas which promise solutions to the problems which underlie their anxieties. For instance, co-author Schlemmer has had the opportunity to review a systematic evaluation of the very dramatic effect on audiences of presentations of Kendall and Louw's *South Africa : The Solution* (which presents maximal political decentralization, devolution and 'cantonization' as a solution to South Africa's problems).

We suggest that the effective *promotion of ideas* for political resolution, directed at support groups of government, is likely to have considerable effect. Obviously the ideas so promoted must persuasively address the powerful latent or manifest anxieties referred to above.

All the processes or potential processes referred to above are examples of deliberate action for change, or what is sometimes referred to as 'voluntaristic' change. Underlying these deliberate strategies are processes of social change which occur without deliberate intervention, which the sociologist, Anthony Smith, called 'intransitive' change.[54] These may be demographic factors, changing values in important political institutions, economic pressures and needs, and management problems which have to be solved by adaptations in day to day government administration.

The latter processes are very pervasive and are likely to qualify, negate or facilitate deliberate change initiatives. Hence, it has become easier to campaign against the Group Areas Act in a situation in which housing shortages have forced many black, coloured and Indian people to move into white areas in defiance of the Act. Up to 200 000 people of colour have moved into white areas according to some recent press estimates.[55]

As we pointed out in Chapter 4, the most powerful source of non-voluntaristic change has undoubtedly been the economy which generated a need for more skills than whites could provide. Its expansion facilitated the emergence of a black labour movement and through inflation and resulting labour unrest, generated a need for change in the labour laws and the recognition of black trade unions in the search for stability. The effects upon a whole range of institutions, including education, state and private manpower-training organizations, and the elimination of certain statutes, have been significant.

The effects of socio-economic black development on political and social change are powerful. As we have suggested in Part One, it was relatively easy for South African white society to justify apartheid when the vast majority of blacks were less educated and occupied unskilled jobs. Now that the number of black people graduating from high school exceeds the number of whites, and blacks are rising into ever more prominent positions in both the public and private sectors, whites have a different conception of South African society.

The government, too, has begun to undermine its own laws. In 1987, for example, 90 per cent of the applications received from Africans for permits to live in white areas were granted,[56] and in 1988 more than half of such applications from all black groups were acceded to.

This brief discussion leads us to a few broad conclusions. South Africa is changing in ways which cannot be reversed and which are difficult to curb. Many of the changes may not be perceived to be significant because they do not address the core issues of political inequality. Yet each change which occurs brings the system of interacting pressures closer to the core issue. In seeking to facilitate this process, popular mobilization by black communities on specific issues, supported by organizations of status and functional importance, such as the business community, will have effects. These effects will occur sooner if the targeted structures are already under pressure because of involuntary change processes.

Each new shift in this process of change erodes the common earlier conviction held by whites that they constituted a separate sub-nation in a multi-national territory. Today, the effective debate is between multi-racialism (or the institutionalized ethnic variety) and non-racialism.   The white centre of gravity is presently situated within the multi-racial model.   Even the conservatives, who would like to cling to the idea of a separate white nation, are now realizing more and more that only partition of a radical (and unrealistic) kind can preserve a white nation in the face of steadily advancing multi-racial interdependence.

In the major debate, however, the issue of how power is to be allocated and distributed is becoming more and more clearly crystallized as socio-economic change strips away the vestiges of social apartheid which, in the past, have 'cushioned' the issue of power.   The quality and consequences of this debate hold the key to the future direction and tempo of change in South Africa.

The processes of change suggested above are all incremental, and while effective, cannot at this stage bridge the gulf of alienation which exists between white society and the black intelligentsia.   The latter feel deprived of recognition, dignity and status, and as long as racial discrimination exists, they will experience a sense of affront.   Indeed, as the various forms of discrimination weaken, so the ones that remain will seem more incongruous and aggravating.   Not even US society has succeeded in removing the sense of affront which black élites feel.

Only effective participation in power on all levels can address the alienation of the black élite.   This alienation is seen clearly in the general refusal to participate in government reform and negotiation structures.   Hence reform and incremental change are steadily increasing the pressures for the political participation of the majority, particularly on the national level.

Before turning to this debate and making our own proposals, we briefly review the prospects for further incremental change over the next five years.   Developments seem likely to occur which will influence the way power is perceived and defended in the years that lie ahead.   Proposals for a resolution of conflict have to be located within an anticipation of what the short-term future holds.

Notes

1   Zach de Beer, 'Fusion', *Leadership South Africa*, vol. 8, no. 3, March 1989, pp. 71 – 2.
2   Ellen Hellmann, *In Defence of a Shared Society.*   Johannesburg: South African Institute of Race Relations, 1956, p. 1.
3   Report of the Political Commission of the Study Project on Christianity in Apartheid Society (Spro-cas Report), *South Africa's Political Alternatives.* Johannesburg: Spro-cas Publication, 1973.
4   Benyamin Neuberger, *National Self-Determination in Post-Colonial Africa.* Boulder: Lynne Rienner, 1986, pp. 12 – 13, 41.

5   Spro-cas, *Political Alternatives*, pp. 243 – 4.
6   F. van Zyl Slabbert and David Welsh, *South Africa's Options : Strategies for Sharing Power.* Cape Town: David Philip, 1979, p. 153.
7   Slabbert and Welsh, *South Africa's Options*, p. 163; see also pp. 39 – 42, 59 – 62, 152 – 3.
8   Slabbert and Welsh, *South Africa Options*, pp. 35, 159. For a critique see Andre du Toit, 'On South Africa's options', *Social Dynamics*, vol. 5, no. 2 1979, p. 61.
9   Joseph Rothschild, *Ethnopolitics.* New York: University of Columbia Press, 1981, p. 62.
10 Walker Connor, 'Nation-building or Nation-destroying', *World Politics*, 24, 1972, pp. 342 – 3.
11 Ed Cairns, 'Intergroup conflict in Northern Ireland', Henri Tajfel, (ed.), *Social Identity and Intergroup Relations.* Cambridge: Cambridge University Press, 1982, p. 277.
12 Donald L. Horowitz, *Ethnic Groups in Conflict.* Berkeley: Universite of California Press, 1985, p. 104.
13 Joseph Rothschild, *Ethnopolitics.* p. 133.
14 Lawrence Schlemmer, 'Social implications of constitutional alternatives in South Africa', in John Benyon, (ed.), *Constitutional Change in South Africa.* Pietermaritzburg: University of Natal Press, 1978, pp. 267 – 8.
15 Lawrence Schlemmer, 'English-speaking South Africa today' in A. de Villiers (ed.), *English-Speaking South Africa Today.* Cape Town: Oxford University Press, 1976, pp. 91 – 136.
16 Lawrence Schlemmer, 'South Africa's National Party Government', in Peter L. Berger and Bobby Godsell, (eds.), *A Future South Africa : Visions, Strategies and Realities.* Cape Town: Human and Rousseau, 1988, p. 27.
17 Robin Wilson, 'Moving on : a new politics', *Fortnight* (Belfast), no. 256, November 1987, pp. 21 – 3.
18 Paul Sites, *Control and Constraint : An introduction to Sociology.* New York: Dunellen, 1973, pp. 73 – 6.
19 Jannie Gagiano, '"Meanwhile back on the boereplaas": student attitudes to political protest and political systems' legitimacy at Stellenbosch University', *Politikon*, vol. 2, no. 2, 1986, pp. 3 – 23.
20 Slabbert and Welsh, *South Africa's Options*, pp. 90, 170.
21 Du Toit, 'On South Africa's options', p. 61.
22 Martin Staniland, 'Democracy and ethnocentism', Patrick Chabal, (ed.), *Political Domination in Africa.* Cambridge: Cambridge University Press, 1986, p. 57.
23 Patrick Chabal, 'Introduction : Thinking about politics in Africa', in *Political Domination in Africa*, pp. 8 – 9.
24 J. K. MacNamara and L. Schlemmer, *Black employees attitudes to intergroup violence on gold mines.* Confidential Report to Industry, 15 April 1988.
25 *Financial Mail*, 31 January 1986.
26 H. Puhe and K. P. Schöppner, *Views of Coalminers in South Africa on Sanctions.* Bonn: Deutsche Afrika Stiftung Publication Series, no. 45, 1987.
27 Puhe and Schöppner, *Views of Coalminers.*
28 L. Schlemmer, 'Data collection among black people in South Africa'. *HSRC Research Bulletin*, vol. 13, no. 1, 1983.
29 Schlemmer, 'Data collection among black people'.
30 Deneys Schreiner, et. al., *The Buthelezi Commission.* Durban: H and H Publications, 1982, Section 4. See also Schlemmer, 'Data collection among black people'.
31 *Business Day*, 21 June 1988.

32   The Report of the Study Commission on US Policy towards Southern Africa, *South Africa : Time Running Out*. Berkeley: University of California Press, 1981.

33   Robert I. Rotberg, *Towards a Certain Future*. Cape Town: David Philip, 1981.

34   Richard Leonard, *South Africa at War : White power and the Crisis in South Africa*. Westport: Lawrence Hill, 1983.

35   Robin Cohen, *Endgame in South Africa*. London: James Currey, 1986.

36   Theda Skocpol, *States and Social Revolution : A Comparative Analysis of France, Russia and China*. Cambridge: Cambridge University Press, 1979, p. 16.

37   *Financial Mail*, 10 March 1989, p. 27.

38   Schlemmer, 'South Africa's National Party Government', pp. 31 – 2.

39   Cited by Deon Geldenhuys, *Some Foreign Policy Implications of South Africa's Total National Strategy*. Braamfontein: SA Institute of International Affairs, 1981, p. 3.

40   Samuel P. Huntington, 'Reform and stability in a modernizing multi-ethnic society', *Politikon*, vol. 8, 1981, p. 11.

41   Stanley Uys, 'Blacks know they are going to win', *The Guardian*, 1 June 1986, p. 7.

42   Jannie Gagiano, 'The scope of regime support : a case study', in Hermann Giliomee and Lawrence Schlemmer, (eds.), *Negotiating South Africa's Future*. Johannesburg: Southern Publishers, 1989, pp. 52 – 62.

43   See for instance the article by Patrick Lawrence in *The Star*, 21 May 1988.

44   John Kenneth Galbraith, *The Anatomy of Power*. London: Hamish Hamilton, 1983, pp. 187 – 8.

45   Merle Lipton, *Sanctions and South Africa : The Dynamics of Economic Isolation*. The Economist Intelligence Unit, Report 1119, 1988.

46   See the articles in *Indicator South Africa*, vol. 4, no. 2, 1986.

47   Puhe and Schöppner, *Views of Coalminers in South Africa on Sanctions*.

48   R. W. Johnson, 'How long will South Africa survive?' *Die Suid-Afrikaan*, no. 19, Feb. 1989, p. 17.

49   Much of the comment which follows is drawn from the valuable article by Robert Jaster, 'The rocky road to Lancaster House : lessons from the Rhodesian conflict', *South Africa International*, vol. 18, no. 2, 1987, p. 107 – 29.

50   Steven Friedman, *Reform Revisted*. Johannesburg: SA Institute of Race Relations, 1988.

51   Steven Friedman, *Understanding Reform*, Johannesburg: SA Institute of Race Relations, 1986.

52   See Chapter 4; see also Brian Pottinger, *The Imperial Presidency*. Johannesburg: Southern, 1989.

53   Among the studies which mobilized information, see Hermann Giliomee and Lawrence Schlemmer, (eds.), *Up Against the Fences : Poverty, Passes and Privilege*. Cape Town: David Philip, 1985. This collection of articles focuses on the detrimental effects of pass laws on the life chances of rural blacks and on sound socio-political development in South Africa.

54   Anthony Smith, *Social Change : Social Theory and Historical Processes*. London: Longman, 1976.

55   L. Schlemmer and L. Stack, *Black, White and Shades of Grey: Residential Segregation in the Pretoria-Witwatersrand Region*. Johannesburg: Centre for Policy Studies, 1989.

56   L. Schlemmer and L. Stack, *Black, White and Shades of Grey*, 1989, ch. 1.

CHAPTER 6

Emerging trends : the future over the short term

Because of the deep divisions and manifestly serious stresses in public life, South Africa creates an impression of unpredictability. Yet the impression is superficial. The basic social structures are deeply institutionalized, as Part One and the previous chapter amply demonstrate. Sudden shifts in societal patterns are not much more likely to occur in South Africa than they are anywhere else in the world. For this reason the probable changes and events over the next few years are unlikely to be very surprising. For the most part they will be the continuation of processes which are already in progress.

Obviously 'wild cards' cannot be excluded. Some international crisis could cause the gold price to sky-rocket; unexpected events could occur in the independence process in Namibia which might affect South Africa dramatically; Nelson Mandela might pass away or might be released and make a dramatic return to the political stage, setting in motion a self-reinforcing chain of events.

This type of event is beyond prediction, however, and all one can do is acknowledge the possibility of the unexpected. For the rest, assessments of the future have to be based on probabilities, not possibilities, and the probabilities have already been signalled.

This does not mean that the future has to be seen as a completely linear extension of the present. Different combinations of events can be mutually reinforcing to produce quite widely varying future scenarios. Beyond the very short term, therefore, the future has to be described in terms of alternative socio-political worlds which can be quite distinctive. For the moment, however, our task is to review the emerging short-term future and the trends which seem likely to produce it.

The economy

While a detailed economic analysis is beyond our scope, a brief view of the economy is necessary as a background to some anticipated socio-political developments. We shall not comment on economic policy as such, but address the larger political question of whether or not the economy is likely to collapse or sink into a severe crisis.

It is generally agreed that the South African economy has fundamental weaknesses and that economic growth is curbed by a balance of payments problem, aggravated by international sanctions on loan finance and external investment capital.

The effect of sanctions must not, however, be over-emphasized. To a substantial degree, the flow of external capital in the period 1985 – 7 was inhibited by perceptions of instability and poor investor confidence in South Africa. These perceptions were based partly on social unrest, partly on what appeared to be mounting levels of labour unrest, an increasing brain drain and erratic behaviour on the part of government.

Business confidence was restored by the economic recovery in 1987 – 8, and a fall in the rate of inflation. This has been reinforced by the decline in the number of mass-based challenges to the state, by the recent rise in the number of immigrants and tourists from abroad, and the imposition of consistent, albeit highly unpleasant, security management by government.

Most recent signs have indicated that the much-anticipated downturn in the economy in 1989 – 90 is not likely to be as sharp as was originally feared. Most estimates suggest a growth-rate slightly in excess of 2 per cent in 1989, a view endorsed by the Governor of the Reserve Bank. This general trend is accompanied by very favourable property prices, even a slight property boom,[1] and the first signs of some renewed investor interest from abroad.

Adding to the emerging, cautious economic optimism is a realization that the informal sector has much more vitality than anyone expected five years ago. South Africa's current rate of real growth may be significantly higher than official figures suggest, even though central economic planners do take some account of the informal sector. Estimates of the size of the informal economy range between 15 and 40 per cent of the recorded economy.[2] Even the more probable lower figure is substantial.

Most significant, perhaps, is the fact that the economic recovery in 1987 and 1988 actually reduced the overall black unemployment rate slightly (as measured by the Current Population Survey which, although it does have methodological limitations, offers some basis for assessing trends).

The large outstanding foreign debt and the fact that the country is committed to massive repayments from 1990 onwards has contributed to the economic uncertainty. If the current phase of economic stability continues, however, it is entirely possible that further deferment of loan repayments will be fairly readily accepted by the consortium of international banks dealing with the debt repayments. Even if the money has to be paid, however, the economy will be severely constrained but is unlikely to collapse.

Taken together, these trends suggest that with the next upswing in the business cycle, real growth could rise to well above 3 per cent per annum. This trend will obviously be self-reinforcing and contribute to a further strengthening of economic confidence. Although the economy has certain fundamental weaknesses (a shortage of higher-level skills, a sluggish rise in productivity, a lack of international competitiveness of exports, a low propensity for internal savings, and a probable rise in inflation in 1989 and 1990) it will be healthy enough to begin to generate sustained, if small, real increases in average incomes once again.

Political developments — parliamentary politics

In the recent municipal elections the Conservative Party took control of a majority of white municipalities in the Transvaal. Nevertheless, its performance was not as good as most observers expected. Based on recent results, it would not win more than 40 to 50 seats (or less than a third of the seats) in the House of Assembly. Furthermore, the effects of its control of a large number of municipalities, highlighted by the furore over the reintroduction of petty apartheid in towns such as Boksburg and Carletonville, will demonstrate the impracticality of many of its policies, and hence its performance in a general election may be weaker than expected.

It is becoming increasingly clear that the rapid rise of the right-wing was, in part, a response to the social unrest and insurgency during the 1985 – 7 period. It was the conservative equivalent of the emigration of upper-middle class liberal whites. Under conditions of lowered basic anxieties, therefore, a future election should confirm the view that the Conservative Party will be a strong opposition but is unlikely, under present conditions of relative stability, to challenge the electoral dominance of the National Party.

Table 6.1 indicates the current state of white electoral politics. The first table gives the total number of seats held by the different parties after the 1987 election, their percentage share of the vote and the number of seats they would capture under a system of proportional representation. The second table gives the position at the end of 1988, taking into account the municipal election of 1988 and attitude surveys undertaken towards the end of the year. The third and fourth tables give the distribution of voters among parties and within provinces in 1987 and 1988.

TABLE 6.1: ESTIMATED OUTCOME OF A HYPOTHETICAL WHITE GENERAL ELECTION. BASED ON A COMPARISON OF THE 1987 GENERAL ELECTION AND THE OCTOBER 1988 MUNICIPAL ELECTIONS

1987 : ACTUAL PARLIAMENTARY ELECTION RESULTS — NUMBER OF SEATS

Party	Before	After	Share of total	Seats as per proportional representation
NP	114	122	51,17	85
PFP	25	20	15,65	26
CP/HNP	20	22	29,20	48
Ind etc.	7	2	3,98	7
Totals	166	166	100,00	166

1988 : HYPOTHETICAL PARLIAMENTARY ELECTION RESULTS — NUMBER OF SEATS

Party	Before	After	Share of total	Seats as per proportional representation
NP	122	105	50,74	84
PFP (now DP)	20	17	16,86	28
CP/HNP	22	44	32,40	54
Ind etc.	2	–	–	–
Totals	166	166	100,00	166

1987 : ACTUAL PARLIAMENTARY ELECTION RESULTS — NUMBER OF VOTES (000's)

Party	Cape Province	Transvaal	OFS	Natal	Total
NP	330	527	93	119	1069
PFP	133	124	3	67	327
CP/HNP	96	417	71	26	610
Ind etc.	30	25	1	27	83
Total votes	589	1093	168	239	2089
No. reg. voters	843	1616	234	360	3053
% Poll	69,34	67,09	72,91	66,22	68,89

1988 : HYPOTHETICAL PARLIAMENTARY ELECTION RESULTS — NUMBER OF VOTES (000's)

Party	Cape Province	Transvaal	OFS	Natal	Total
NP	362	489	88	121	1060
DP	119	134	4	95	352
CP/HNP	108	470	76	23	677
Ind etc.	–	–	–	–	–
Total votes	589	1093	168	239	2089
No. reg. voters	843	1616	234	360	3053
% Poll	69,34	67,09	72,91	66,22	68,89

Sources:

D. J. Van Vuuren; J. Latakgomo, H. C. Marais and L. Schlemmer, (eds.), *South African Election 1987*. Durban : Owen Burgess Publishers and Centre for Policy Studies, 1987 .

Market and Opinion Surveys (Pty.) Ltd, Bellville, Cape.

The estimates were made jointly by R. Bethlehem and L. Schlemmer, taking as basis the 1987 election results, the recent municipal elections and the results of national opinion polls.

From now on attention is likely to shift to the new Democratic Party (DP). Opinion polls conducted before the merger[3] suggested that the combined strength of its three constituent groupings would be some 24 per cent of the total white electorate. The assumption among party personnel is that it would gain up to 30 per cent of the vote in a general election. These expectations have been strengthened by municipal by-election results in Johannesburg and Port Elizabeth.

Nearly one half of the hypothetical supporters of the new DP are substantially more conservative than the typical support group of the PFP and the National Democratic Movement (NDM). If the new party is not able to project at least part of the agenda and image of Denis Worrall's Independent Party (IP) grouping, voters will either abstain or defect to the National Party. The DP is going to face formidable difficulties in coping with the requirements of the wide spectrum of white voters which it has targeted as its support group.

There does not appear to be a leader on the horizon who could effectively unify the different strands of ideology within the liberal opposition. Table 6.2 demonstrates the differing opinions on some crucial issues of change between the parties which together comprise the DP. There are very marked differences in attitudes between the IP supporters and the PFP/ NDM supporters on key issues of security and negotiation. It is possible that the DP will somehow bridge its internal differences and gain more seats than Table 6.1 suggests, and in splitting the NP-vote, increase the CP's victories. The NP is likely to retain an overall majority, however.

TABLE 6.2: RESULTS OF A SURVEY OF WHITE SUPPORTERS OF DIFFERENT POLITICAL PARTIES REFLECTING OPINIONS ON INTENSIFIED SECURITY ACTION AGAINST ACTS OF TERRORISM AND NEGOTIATIONS WITH THE ANC. (Expressed as percentages).

	CP	NP	IP	NDM	PFP	Total
Intensified action						
against ANC – Important	93	93	79	56	52	85
– Unimportant	2	–	7	27	29	6
Negotiations with the						
ANC – Important	16	23	57	87	75	34
– Unimportant	65	54	17	2	9	44

Note: The balance were either neutral or gave no answer.

Source: National sample of 1630 white voters; survey conducted by Market and Opinion Surveys (Pty.) Ltd, November 1988.

Comment: These results show that IP (Worrall) supporters tend to lie between the NP and the PFP/ NDM supporters in key political opinions.

Over the short term, dominant action in Parliament is likely to be the contest between the Labour Party (which has an unchallenged majority in the House of Representatives) and the government over the tempo of reform. Although the Carter Ebrahim faction is likely to become more vocal in its opposition in the House of Representatives, the position of the Labour Party is probably unassailable.

The outcome of this conflict cannot be predicted. However it is likely that the stand-off or stalemate in the contest will continue with some waxing and waning of rhetorical intensity until such time as the government is ready to abolish the Group Areas and Separate Amenities Acts and is prepared to soften its position on segregated, white, government schools, perhaps by allowing a form of local option. A more conciliatory leader than P. W. Botha, able to negotiate and 'horse-trade', may be able to strike a deal with the Labour Party over the pace and pattern of reform, thus resolving the present stalemate in Parliament. The new NP leader, F. W. de Klerk, is promising in this respect.

Whatever the case, the Labour Party will be the dominant 'opposition' within South African parliamentary politics for the foreseeable future in view of its veto right over constitutional change. One might expect a few defections from the Labour Party, perhaps as a result of rewards and blandishments offered by the government, but present indications are that the most competent members of the Labour Party caucus will remain united.

In a general election one would expect the National People's Party to suffer a setback in the House of Delegates, as a result of a recent scandal surrounding its leader Amichand Rajbansi. The Solidarity Party will probably gain the ascendancy and the Labour Party could anticipate some cautious support from the House of Delegates. Generally speaking, however, the House of Delegates tends to be conservative in its stance and will draw back from high-key confrontation with the National Party.

Extra-parliamentary politics and other black politics

There seems to be general agreement that the State of Emergency has placed a damper on the activities of the extra-parliamentary organizations. Detentions and the threat of detentions have severely constrained the leadership and hence also the activities of those organizations linked to the UDF.

In some of the major black townships, however, there is still a determined and coherent core of activists. These groupings enjoy links with organizations in the white professional, academic and 'progressive' communities as well as with the embattled but energetic alternative press. There is also an increasingly active interchange between internal networks of radically-orientated blacks and the ANC in exile, seen recently in the form of visits to Lusaka by trade unionists, professionals, businessmen and managers.

The existence of an effective stalemate resulting from the State of Emergency has produced a debate within the network of extra-parliamentary organizations on the merits and demerits of contact with or participation in official structures.[4] This debate is not likely to result in a unified new strategic direction in the short term, however, mainly because of the alienation referred to in the last chapter and a concern among ANC leaders that its overall strategic coherence and internal allies could fragment if such participation were to occur.

Nevertheless, selective negotiation or interaction with the authorities around specific issues and problems is likely to occur on the Witwatersrand. Elsewhere, for a range of specific reasons, a more hostile attitude to any form of state authority is likely to continue.

In Natal, the energy of the extra-parliamentary movement has been dissipated by the ongoing conflict and violence between lower levels of the UDF and Inkatha — in most cases the groupings actively involved in the conflict appear to have rather ill-defined links to the echelons of leadership in the two organizations.

While it is possible that some kind of truce between the two organizations will be hammered out or will emerge because of mutual exhaustion, the antagonisms at grassroots level are likely to subside more slowly, as the mutual destruction has taken on the characteristics of a feud, with vendetta-based action and counter-reaction. It is probable, however, that this ongoing conflict is weakening the popular support-base of both Inkatha and the UDF in Natal, as rank-and-file people find their lives disrupted to no particular end.

In general, the lack of strategic success experienced by all of the black community-oriented extra-parliamentary organizations will weaken grassroots following and a vacuum as regards political mobilization and popular involvement at rank-and-file level will continue for the foreseeable future. Extra-parliamentary politics will depend for its survival on the professional and intellectual networks referred to earlier.

There is a well-developed political 'vanguard' in the extra-parliamentary organization but a rather fragmented following. Furthermore, the combined effects of the State of Emergency and the tendency on the part of the spokespeople to defer to imprisoned or exiled leaders have resulted in a situation in which no internal leadership with a large popular grassroots following exists. Prominent churchmen such as Archbishop Tutu fill the gap, as it were.

Black politics among the 'participating' groupings is likely to be more varied. The clear, and repeatedly signalled, desire on the part of government to engage in negotiations with black organizations will tend to increase the confidence of the new black local councils and some fairly interesting leadership is bound to emerge. For example, the Sofasonke Party which controls the Soweto City Council under Mayor Makhwanazi is emphasizing grassroots needs and grievances in a bid to increase its popular following.

It is also probable that the government will have a measure of success with its plans to establish a series of regional bodies to represent and negotiate for black local communities, using local councils as an electoral college to elect regional representatives. All this will occur in an atmosphere of heated controversy and criticism by extra-parliamentary groups.

It will be enormously difficult to manage what will be a complex and

new process, and the ultimate prospects of establishing fully functioning and reasonably legitimate black institutions at regional level are
probably very mixed at this stage. As we will argue later, major shifts
in 'intention' are required from both the government and the ANC
before an appropriate climate will be created which will facilitate this
kind of 'bottom-up' development of black political participation.

Turning to the exiled movements, the ANC will continue, on the one
hand, with its rather mixed and ambiguous strategy of attempting to
win hearts and minds within South Africa by an ongoing series of
meetings with visitors from South Africa, while at the same time
symbolically demonstrating its relevance and power through insurgency and sabotage.

Of more relevance, perhaps, will be the debate around the new
constitutional guidelines of the ANC. These guidelines include endorsement of a multi-party democracy, a bill of individual rights and a
mixed or social democratic form of economic organization. While
rather vague at this stage, these guidelines will raise the ANC's profile
with Western governments. They appear to have been produced partly
in response to questions about the organization's strategy of violence,
raised by the former US Secretary of State, George Schultz, and by the
British government. Over the short term one might expect certain
Western governments, possibly supported by the USSR, to keep up
pressure on the ANC to play down its military struggle in favour of
initiatives more appropriate to eventual political negotiations with the
South African government.

Prospects for socio-political reform

The tempo of reform depends quite considerably on how soon President
Botha's successor can consolidate his position within the party and the
executive branch of government. Mr F. W. de Klerk's speech to
Parliament, immediately after assuming the leadership of the National
Party, did not signal any major changes in specific government policies
but it was considerably less ambiguous than previous general policy
statements made by P. W. Botha. He promised a more energetic and
positive approach to both reform and negotiation in the future and
committed himself to strive towards the goal of achieving majority
support for the NP among all South Africa's ethnic groups.[5]

This may appear to be a very tall order, but it is an attempt to position
the NP as a potential multi-ethnic party which could unify moderate
conservatives across the spectrum of the classified groups. It is perhaps

of interest to note in this regard that in March 1989, an opinion poll conducted by the British newspaper, *The Independent*, through the SA research organization, Markinor, suggested that Mr P. W. Botha was second in popularity as a future State President among urban Africans. Nelson Mandela was chosen by 41 per cent of respondents, while P. W. Botha and F. W. de Klerk were selected by a combined 22 per cent. Hence, the idea of the NP as the core of a substantial multi-racial conservative alliance may not be far-fetched.[6]

Broadly speaking one could expect the following developments as regards reform over the next two years:

- The Group Areas dispensation is likely to become substantially modified primarily by two processes. One will be a steady erosion through a process of informal 'greying' which the government will knowingly tolerate in areas where there is no mobilized white resistance, and an expansion of the number of areas formally declared to be Free Settlement Areas. Although this process will be centrally managed, the selection of areas might involve some kind of local option built into the process of decision-making. One might expect some peri-urban areas to be 'released' for this process by offers of alternative housing to the local white residents who do not want to remain. Attempts will be made to secure the co-operation of the Labour Party in this process through the parliamentary joint committee system.[7]

- Certain white government high schools (particularly those in upper-middle class, English-speaking pupil catchment areas, and under-utilized, centrally-located establishments) are likely to be allowed to admit a *quota* of black, coloured and Indian pupils. The quota system, as was the case with universities, will essentially be a transition mechanism. At some point some of these schools may be given a new 'constitutional' classification, taking them out of the 'own affairs' domain, which will open the way for a more rapid enrolment of black pupils and a general move toward open state schools. An alternative would be to privatize these schools and allow them to function as subsidized multi-racial institutions.

- The political dynamic surrounding the Separate Amenities Act will induce NP-controlled municipalities to move faster in declaring amenities open to all, which will pave the way for the lifting of the legislation. The government has already signalled this by instructing police not to arrest blacks found at white facilities but to refer cases to the Attorney-General[8] who is unlikely to prosecute many cases. The government is also considering legislation to prevent facilities which have already been opened to all races from being closed by Conservative Party city councils.

- One can expect an increased tempo in the release of land for site-and-service housing in peri-urban areas. Urban guide-plans will be modified in order to open new areas for formal black township development. Nevertheless, controversies surrounding shack settlements are likely to continue as a result of the huge current housing backlog of approximately one million units (estimate by the Urban Foundation).

- As indicated earlier, there is a steadily growing prospect of the KwaZulu-Natal Indaba being resurrected as a negotiating issue between government and Chief Buthelezi. One cannot expect immediate results but it is likely that, within two years, some form of new proposals will emerge from government for the unification of the two regions. It is uncertain, however, whether agreement between the central and KwaZulu governments will follow upon this.

- There will be increasing speculation about the prospect of the government moving away from a formally-defined group-based society to one based on voluntary association. Already there is considerable speculation as regards the implications of Free Settlement Areas, centred around the question of whether or not they herald a shift towards non-racialism in the constitution. This is likely to be a long drawn-out issue but it was placed on the agenda by Chris Heunis in his role as Acting State President.[9] Within the next two years we are likely to see attempts to accommodate a fifth statutory group — an 'open' group.

- In the meantime, however, there is every likelihood that the National Council (the government's proposed forum for negotiation with Africans) will be established, amid great stress and controversy. The establishment of this body is likely to be destabilizing to an extent, since there will be redoubled efforts on the part of extra-parliamentary groupings to challenge the legitimacy of these developments. However the blanket impact of the State of Emergency makes it unlikely that the unrest will reach the proportions evident in the 1984 – 6 period.

- While these developments are occurring, substantial use will be made of methods of co-option in an attempt to draw more and more Africans onto decision-making bodies. The government will continue to attempt to circumvent the Labour Party veto on constitutional changes. If it succeeds, Africans will be appointed to the cabinet.

- As part of the process of establishing African regional authorities as feeder-institutions for the National Council, consideration may be given to sub-dividing the provinces into smaller constitutional units to coincide with the nine planning regions. The next two years could also see the re-introduction of elected white, coloured and Indian representation (along group lines) at second tier level, to which African representation will be added within the regions mentioned above.

- The government might also attempt to unify Parliament by establishing an over-arching single chamber, albeit based on formal racial component-houses.

Over-arching and external factors

The probabilities outlined above spell out a complex, slow and controversial process of socio-political adjustment. At no stage will observers (or participants) develop a clear feeling of breakthrough or of a major resolution of the country's problems. Nevertheless, a new type of interaction between political communities will be developing which could steadily strengthen the stability of the 'system' in the medium term. This could also set the stage for an evolutionary process of change towards what could eventually become a form of multi-racial semi-democracy, incorporating federal elements based both on group rights and regional representation.

On the basis of the scenario outlined, the social and political system will be mutating steadily, in a sense preparing the country for the more far-reaching changes that are necessary for a resolution. During this process the right-wing opposition will read the signs with despair and warn *ad nauseam* of an imminent sell-out of white interests.

The liberal opposition will claim, equally predictably, that too little change is occurring. Although neither argument will be convincing enough to capture substantial additional support among white voters, these half-truths could lend credibility to the activities and programmes of both right- and left-wing radicals.

In order to succeed, this process will require very skilful security management and a great emphasis on internal diplomacy. There is abundant evidence that the government has become more alert to the effects of its actions than it was previously. This learning curve is likely to be sustained.

The kind of 'evolutionary fluidity' we may be entering existed, to an extent, in the early 1980s — the period preceding the Tricameral Parliament — but a series of developments caused the entire process to collapse into prolonged instability. What are the prospects of this occurring again?

As we indicated at the beginning of the previous chapter, the 'collapse' which occurred in 1984 involved a coincidence of events which are unlikely to be repeated. There are no immediate and powerful internal threats to what might be called the 'revolutionary stalemate' in the South African political economy which exists at present.

However, things can deteriorate rapidly. This can happen in a variety of situations, for instance, if very damaging international sanctions are imposed; if as a result of ANC activity in the Frontline States there is a return to the politics of regional destabilization; if the Namibian independence process experiences serious crises, or if the release of Nelson Mandela triggers an exceptionally powerful wave of internal unrest. In such instances activists will, in a sense, be 'protected' against security action by external pressures and expectations. Any of these developments could undermine the fragile recovery of economic confidence, which could, in turn, have powerful political consequences and possibly lead to sustained instability.

For this reason one has to take account of the possibility of what may be termed a 'downside' scenario, in which:

• internal protest will well up again;
• the right-wing will regain the ascendancy;
• the economy will suffer, deepening both the above developments;
• external pressure will increase.

The end results of such a scenario are unpredictable in that they could resolve themselves in more than one way.

Nevertheless, in its new crisis-management mode, the government is likely to try and avoid stimulating the kind of damaging processes referred to above. Although the downside scenario cannot be excluded, the prospects of it emerging are not high.

In the longer term, new challenges and difficulties will undoubtedly arise which will introduce different and perhaps unexpected dynamics. It is for this reason that no sector of the South African community can afford to be complacent.

More particularly, it is appropriate to consider very briefly whether or not the very latest signals of policy changes by government spokesmen hold the prospect of opening the way for negotiations between the government, the ANC and other contending parties.

As already indicated, voluntary group association has been raised by

Minister Stoffel van der Merwe. The possibility of a voluntary 'open' constitutional group has been raised by Minister Heunis: a development which will be given impetus by increasing numbers of officially designated Free Settlement Areas in the major cities.

Just before his resignation in May 1989, Heunis also predicted the introduction of a single legislature to replace the present Tricameral Parliament and a multi-racial executive, which would include African cabinet ministers. While it was widely speculated that Heunis' prediction went somewhat ahead of NP thinking (and may have helped to precipitate his resignation) there are other, less formal indications from parts of the NP caucus that the Tricameral Parliament is being reconsidered or that its future is at least being debated.

The leader of the NP, F. W. de Klerk, while not rejecting these possibilities, has indicated very firmly that no developments being considered will alter the firm commitment of his government to the maintenance of separate power-bases (i.e., racially-defined voters' rolls).[10] This kind of firm qualification by De Klerk was only to be expected. There is, as yet, no indication that the NP will put the coherence of its racial power-base at risk.

Hence, if the speculation about voluntary association and an open group become reality, they will add to or complement the existing racial constituencies (voters' rolls). If the new open constituency is allowed only in districts which have by then become formally desegregated, it will be a small constitutional category. Its very marginality will mean that it will not be taken seriously by the extra-parliamentary movements. It will contribute very little to the kind of climate in which a major resolution could occur. It will be seen as a complex and convoluted way of obscuring the real political issues in the country.

If, however, the voluntary open group is allowed to be established in all parts of the country, it could become a parallel form of political mobilization, coexisting with the racially defined mobilization which exists at present. It could therefore become the constitutional vehicle through which the majority-based non-racial organizations, such as the UDF and the ANC, could enter the formal political system. Even with this kind of constitutional opening, however, the real prospects of resolution and the prospects of attracting widespread participation from extra-parliamentary movements would depend on the arrangements which are agreed upon for the composition of the executive.

Thus, whether or not an 'open' political group is established, a very specific approach will be required at the level of executive power-sharing. We shall discuss an approach which seems to be required in the chapters which follow.

In the following chapter we present a framework which would enable the contending parties to move closer together in the search for a system that would accommodate mutual fears and aspirations.

Notes

1   *The Star*, 25 February 1989.
2   Centre for Policy Studies, *Policy Perspectives: South Africa at the End of the Eighties*, 1989.
3   Market and Opinion Research (Pty.) Ltd, 1989.
4   See Mark Swilling, *Beyond Ungovernability*. Johannesburg: Centre for Policy Studies, 1988.
5   *Die Burger*, 19 February 1989.
6   Partial results of this poll were reported in *The Star*, 30 March 1989. The sample was a carefully stratified selection of 550 black adults in the major metropolitan areas of the PWV, Durban, Port Elizabeth and Cape Town.
7   See L. Schlemmer and L. Stack, *Black, White and Shades of Grey*. Johannesburg: Centre for Policy Studies, 1989.
8   *The Star*, 24 February 1989.
9   *The Beeld*, 10 February 1989.
10 *The Star*, 12 May 1989.

CHAPTER 7

Towards resolution : a framework

Any framework for conciliation must involve two elements. It must have particular features facilitating conflict resolution and it must contain principles of a general nature capable of addressing the major needs and interests within a very wide spectrum of mobilized political groups. We turn first to the more general principles and, in the final chapter, develop these together with particular strategies for political accommodation in South Africa.

Negotiations for an accountable system

Although a negotiated settlement approaches the ideal shared by most South Africans across colour lines, this option is not the first choice of either the current NP or ANC leadership. Both groups insist on their own particular political solutions and their own conceptions of the South African nation. The NP's notion of political legitimacy is an ethnic power balance, whereas for the ANC legitimacy rests on the imperative of numbers or majority rule. Thus to talk about a negotiated accommodation is to talk about compromises and contraints on both sides.

The two main questions which have to be addressed are:
• What is negotiable?
• Is a democratic form of accommodation possible?

As regards the first question, it is important to repeat one of the conclusions reached in Chapter 5, namely that negotiations for a transfer of power to the majority will be fiercely resisted by the NP government and the white community, despite assurances that whites' individual rights and material interests will be protected.

We need to stress again why this is so. In studies of conflict resolution there has latterly been a break with the notion that conflict is fundamentally only about interests. Instead there is a growing realization that the major concern in divided societies is over needs and values such as security and identity, which are extremely difficult, if not impossible, to negotiate.

The new approach to conflict resolution is not to insist on negotiations which solve the problem, but rather to concentrate on facilitating an evolutionary process. This process must promote a greater fulfillment of social needs and enable parties to move deliberately from their present position of conflict to a point in which the security of conflicting parties is mutually enhanced. As Burton concludes, 'In this way change can be more than the mere substitution of one ruling élite for another who will also pursue sectional interests at the expense of human needs'.[1]

As regards the question of whether or not an accommodation can be democratic, we have already argued that the conditions for a liberal democracy in South Africa do not exist, at least not for the time being. However, to say this is not to close off the search for democracy in South Africa. Over the past decade there has been an ongoing, wide-ranging political discussion in South Africa focusing on the feasibility of a specific democratic alternative, namely consociationalism or power-sharing between groups.[2] Societies such as Switzerland, the Netherlands, Belgium and Lebanon have been used as comparative frames of reference.

Participants in this debate have rarely turned their attention to the rather more relevant examples of African countries north of the Limpopo. Like South Africa, most of these countries have had a history of European colonization and are racked by deep ethnic cleavages. Like South Africa, they have to deal with the disastrous combination of severe economic constraints (particularly since 1974) and rapid population growth. Confronted by the collapse of effective government in many parts of Africa, scholars, and in particular Africanists, have been grappling with the crisis of democracy on the continent. This debate is reflected in an important new collection entitled *Political Domination in Africa*.[3]

This collection reminds readers of the appeal made by a doyen of Africanists, Melville Herskovits. He urged Western scholars studying Africa to distinguish clearly between the concept of accountability, which is universal, and the culturally specific forms of it, such as liberal democracy which developed in Western social conditions.[4]

Noting the failure of liberal democracy in Africa, a prominent contributor, Richard Sklar, examines alternative forms of democracy in Africa which also uphold the principle of accountability. There is, for example, *guided democracy* which disposes of the political method of multi-party electoral competition, but insists that rulers should be accountable to their subjects. There is *social democracy* which implies a planned, interventionist pursuit of an egalitarian social order. There is also *participatory democracy* which emphasizes the interaction between political parties and other community-based institutions, particularly democratic trade unions. Sklar proposes as an ideal for Africa, *development democracy*, which is a synthesis of the various forms of democracy.[5]

Although development democracy does away with multi-party electoral competition it must be distinguished quite clearly from dictatorship. In Mobutu's Zaire or Sekou Toure's Guinea, the dictator, in Sklar's words, 'makes no pretence of accountability to the people on the part of an exalted person or national saviour'. Here the single party is, to cite another scholar, 'at the mercy of one man's *diktat* or ... is nothing but the institutional appendix of a clique'.[6]

While developmental democracy might operate in the one-party context it should elevate accountability to a cardinal principle. Accountability entails the following: leaders must have a broad support base; there must be interaction between leaders and the masses; leaders must recognize that power always entails responsibility, something for which they are held accountable; political organization has to be allowed to ensure accountability, and lastly, the government's primary aim should be an improvement in the living conditions of all its citizens. Failure to achieve economic development almost inevitably produces a crisis of accountability.

In short, the overriding aim has to be good government in pursuit of development, an improvement in the quality of life and nation building. In *Political Domination in Africa* it is argued that all this can take place outside the particular framework of liberal democracy. Ultimately, good government is measured not in terms of the liberal democratic credentials of the political system, or the benign intentions of the rulers, but rather in terms of the consequences of their rule for those over whom they rule.[7]

Can a government which is accountable, but not a liberal democracy, produce good government in South Africa? The main stumbling block here is the problem of political sovereignty which is vital for accountability. Without sovereignty, accountability is irrelevant. If the apparent rulers (or some of them) are really the servants of some other power, 'then they can scarcely be held answerable to the people whom they claim, it turns out incorrectly, to rule'.[8]

Political sovereignty is generally believed to reside in the nation. But who comprises the nation in South Africa? We are confronted with the fact that there are two major competing conceptions of the nation in South Africa — a white-led, multi-racial nation under National Party domination which builds on the four statutory groups, or the ANC's conception of an African-led, non-racial nation which demands the elimination of all forms of *political* ethnicity.[9]

Only if there is a prospect of achieving broadly-based agreement on the composition of the 'nation', or if a credible conception of 'dual nationhood' could be developed, does an accountable government become possible. This in turn depends on whether the contending parties can achieve a shared understanding of the balance of power and the limits of their respective political capabilities, and whether they are prepared to accept a framework in which they can accommodate their differences.

As close observers of the conflict in Northern Ireland have come to understand, the contenders in the South African conflict have to realize that the convergence of their separate agendas will, of necessity, be slow. However, progress could be made if each side developed a grasp of what peaceful accommodation requires.[10]

In the long term it is inevitable that South Africa will come under a form of majority rule. Our arguments are in no way intended to challenge or circumvent this ultimate destiny. Neither do they try to suggest an ideal solution. We are concerned with the period of transition — a 'buffer' phase — which South Africa has already entered. The objective of this chapter and further arguments in the final chapter is to suggest a transitional political system that has the capacity of attracting support from both the parliamentary and extra-parliamentary formations.

There is a real danger of the struggle in South Africa degenerating into a debilitating war of attrition in which the only question is who ultimately will inherit the ruins. To avoid this fate, it is fundamentally important that in the transitional period, South Africa must evolve a political system which is able to get the main forces to work together for the common good. Such a system will not come about, however, unless the main contenders mutually accept a framework for co-operation. To our mind, such a framework should include the following elements:[11]

An acceptance of the limits of power

For a development democracy to have any prospect of success in South Africa it is in the interest of both sides to avoid an all-out struggle for national survival on the part of whites, or the national 'reconquest' of the land on the part of the hitherto excluded majority. This means that a revolution, or even a substantial weakening of the state, apart from being unlikely is also undesirable. We explained our own reasons for such an assessment in Chapter 5. Here we cite Dr Mangosuthu Buthelezi who, in 1987, spelt out the lack of success of the armed struggle in brutally clear terms:

> After 25 years of endeavour every bridge in the country is still intact. Every system of electricity and water supply is intact and there is not one single factory out of production because of revolutionary activity. The classical circumstances in which armed struggles win the day include a divided army and security service, a divided civil service and a massive support system on the ground for revolutionary activity. These are just not present in South Africa.[12]

Unlikely though it might be, there are people on the far right and left of the spectrum who would defend the desirability of revolution. We would simply point out that, unlike the case in many Third World dictatorships, the targets of so-called revolution in South Africa are not part of any small, corrupt clique that would be eminently dispensable in the future. The targets on both sides are large and valuable segments of our society. A right-wing revolution would temporarily destroy all majority-based political mobilization. A revolution from the left would attack a large, highly-skilled and experienced echelon of managers, investors, owners and administrators. Can a revolutionary society turn into a development democracy if it has amputated its vital organs? A revolutionary strategy must surely defeat the attainment of mutual trust and nation-building.

On the other hand, it is equally necessary for the state to accept its inability to restore full stability and confidence. South Africa resembles several other divided societies in that a new pattern of conflict seems to have emerged: low-key violence which leads to neither full system breakdown nor to renewed full control by groups that previously dominated almost effortlessly. In the same speech cited above, Buthelezi spelt out the limits of state power:

> The South African government in turn can make no real headway against revolutionaries. If they could have done more to defuse the revolutionary climate they would have

done more. It is still there. There will be no economic recovery unless the confidence factor again plays its definitive role. There will be no return of confidence in a situation in which there is only the prospect of the equilibrium between apartheid and the violence it solicits being raised to ever higher levels. This upward spiralling of violence could stretch into the distant future. We now need real reform to defuse the violent situation.[13]

In an essay on political accountability in Africa, John Lonsdale writes that all rulers hope to convert their force into power, and to carry society with them rather than fight against it. He adds that effective power cannot avoid submitting itself to some test of accountability.[14] Although a government may try to cheat the test or dictate its terms, this becomes increasingly difficult over time, as the South African government has found. It will have to accept that repression cannot restore full stability and confidence. It will have to persuade its adversaries to exchange revolutionary mobilization for parliamentary participation.

In essence, this means that both the parliamentary and extra-parliamentary forces will have to become engaged in politics, as defined in Bernard Crick's *In Defence of Politics*. In the terms of this classic essay, politics means that the tendency to make absolute demands should be abandoned in favour of a realistic conception of what can be gained. This means an acceptance of constraints and a willingness to reconcile different interests.[15]

The acceptance of two political traditions

A contributor to the volume *Political Domination in Africa* writes: 'Nationalism can speak unto nationalism; liberalism often seems just to be talking to itself'. However, he adds that for the former to occur, there must be 'a relatively clear and self-confident identity on both sides'.[16]

Unfortunately the ability of two nationalisms to communicate is seriously impaired if they share the same homeland and are locked in an unresolved struggle. Much of the political debate in South Africa is taken up by the effort to deny or obscure the essential reality of the conflict, namely that it is primarily a struggle between Afrikaner and African nationalists.

Afrikaner nationalists have, until very recently, tended to brand any form of radical black opposition as communist-inspired activists and surrogates of Russia. African nationalists, on the other hand, tend to present the conflict not as a bi-communal struggle for national identity but a struggle of good against evil, adding in the same breath that there can be no compromise with such a pernicious system as apartheid.

However, Afrikaner leaders have always understood that apartheid was merely an instrument of Afrikaner nationalism, never a goal in itself. In the words of Gerrit Viljoen: 'Our policy of separate development is, to my view, no ideology, dogma or principle; it is an instrument or method, a road whose future course displays a certain openness, but which will always have as its goal or future destination the maintenance of our national identity'.[17]

As the odious features of apartheid are stripped away, it will become increasingly difficult to present the struggle as being limited to apartheid issues.[18] Increasingly, the struggle will be over the idea of individual versus group representation and over competing conceptions of nation-building.

The government's cardinal constitutional principle is that the only workable form of representation in a divided society is group representation. Its favoured constitutional solution is a form of corporate federalism, which demands that the various ethnic groups be represented in such a way as to ensure that no single 'nation' dominates government.

The debate between the NP and the Conservative Party (CP) is about the extent to which the demand for white self-determination should be linked to exclusive claims to power, areas of residence and to the land as a whole. The CP demand is for a sub-society with its own national territory, wielding exclusive power. The NP sees whites as a national minority or sub-society which has to develop a form of co-existence with blacks within the same geographic domain. While strongly demanding self-determination for whites, the NP government is prepared to draw Africans (as a corporate group) into government. Although no longer insisting on the exclusive right to shape the constitutional future, the NP will continue to insist on retaining a white veto and white participation.

As far as nation-building is concerned, the NP government is shifting to a new model, that of the South African *state-nation*.[19] In this model, the state, which is becoming increasingly multi-racial, gradually attempts to weld the various cultures and nationalities into a new 'nation-like' entity which will increasingly identify with the state and be prepared to defend it against radical challengers. National Party politicians and civil servants in the executive branch of government, as well as homeland politicians and bureaucracies, are beginning to formulate the idea of a new nation.[20]

A completely different approach to political representation and to nation-building is expressed by the various African movements, ranging from the ANC to the PAC, and from the Black Consciousness Movement (BCM) through to Inkatha. This does not start with the concept of the group or the 'tribe', but with the individual. They believe that political representation must occur on an individual, non-ethnic basis.

Although there are ideological differences within the African movements they all have a common conception of a 'territorial nation' which transcends ethnic divisions. These movements demand that the majority elects the government; this will make the state truly an embodiment of the nation, in other words a *nation-state*. It can be expected that such a state will be strongly African nationalist in its symbols and myths. It will reject the white communal or Afrikaner nationalist claims. It will also set itself firmly against political differentiation between individuals on the basis of race and/or ethnicity, since this would conflict with the majority principle.

African nationalists acknowledge that there is no ready-made South African nation. In a series of papers on the 'National Question' published in *The African Communist*, a contributor states frankly: 'The South African nation is in the process of being born, and we, in the ANC — in an embryonic form — represent the unborn South African nation'.[21]

It is this lack of developed nationhood which has made the issue of political strategy all-important. The ANC is prepared to achieve national self-determination for Africans through alliances with the 'progressive forces' among the other population groups. The PAC and BCM, on the other hand, want to draw only on members of the black groups. They reject alliances, believing that authentic self-determination for blacks cannot be won with the help of whites. However, both the ANC and PAC are nationalist in that they demand the return of the land to its rightful African owners and demand to be governed by their own kith and kin. While Buthelezi's Inkatha movement is obviously influenced by its preponderantly Zulu-speaking base, the organization and its rank-and-file membership, at least in its public pronouncements, rejects 'tribalism' or ethnicity as a goal in itself.[22]

To sum up: the NP's concept of nation-building stresses cultural diversity and multi-nationalism, ideally under white leadership but at least with parity of power for whites. Any African nationalist variant would not countenance white predominance, would promote ethnonational homogeneity, would strongly insist on individual representation, would resist white group representation, and would impart to the state an African character to which all personnel would have to subscribe whether they be Africans or not.

In parenthesis, we acknowledge fully that both the ANC/UDF alliance and the NP have increasingly attempted to transcend the nationalist core identities of their movements. This obscuring of the original core identity has been made necessary by South Africa's powerful modern industrial economy which has enforced the interdependence of various groups. As Afrikaner capitalism has grown and interacted with English-speaking capital it has become functional for Afrikaner nationalism to draw into its circle of mobilization the English-speaking middle class. The CP has followed the same strategy.

The ANC, on the other hand, recognizing that part of its target for attack is the white economic establishment which coexists with Afrikaner power, has formed an equally functional alliance with socialists, communists and humanist anti-capitalists of all races. Both the NP and the ANC are hence *alliance* formations. To protect and promote their respective functional alliances, both movements have had to play down their original core identities. These identities remain very salient, however. To those who deny this we pose a very simple question: could a white or an Indian become the president and symbolic leader of the ANC? Could a white without Afrikaner roots or credentials become leader of the NP? The answer in both cases is very definitely no!

Is there any hope of a compromise? Perhaps the conflict is not as stark as it seems. It is true that most whites insist on control over their own destiny through representation of whites by whites; however, NP caucus and cabinet members, in private interviews, no longer demand that each of the other statutory groups need only be represented through their own statutory group. The NP may be prepared to negotiate on the issue of participation by an 'open' category of people. It may also, over the medium term, allow the formation of a broadly-based non-racial party.

The position of blacks on this point is also ambivalent. Recent research into popular attitudes among Africans suggests[23] that the rank-and-file would accept, indeed possibly even prefer, forms of balanced power-sharing between groups, as opposed to majority domination. For example, in 1987 some 75 per cent of a representative sample of Africans in the Pretoria-Witwatersrand-Vereeniging (PWV) area, regarded a multi-racial government of Africans, whites, coloureds and Indians, without a group domination, as 'good' or 'excellent' compared with only 45 per cent giving a similar evaluation to an African-dominated government.[24] In a more recent survey, only 33 per cent of Africans in the PWV endorsed a majority rule dispensation compared with various group power-sharing options.[25] The results of the polls done on attitudes to ethnicity in politics, presented in Chapter 6, are also apposite.

On the other hand, the current ANC leadership strongly rejects any accommodation of ethnicity, its earlier participation in the ethnically-structured Congress Alliance notwithstanding. Oliver Tambo has called a system of white minority rights racist and anti-democratic, since it would build on the thesis that whites are different from and opposed to the rest of the population.[26]

Nevertheless, this may be an issue open to compromise. Obviously Africans (and most blacks) would be opposed to a white group choosing its own (white) representatives if they were to form part of an overall system of white domination. However, should the members of the black groups be allowed to decide on their own form of representation within a common power-sharing government, the extra-parliamentary opposition may well be prepared to accept white representation through a party which is predominantly white and enjoys the confidence of whites.

In theory, then, a compromise may be possible between a white demand that their own trusted representatives speak on their behalf at the negotiation table or in government and the African demand for colour-blind or individual representation. The main qualification would be that every person should be free to associate with the party of his/her choice, and that the system should not be a cover for continued political domination.

Among political theorists there is by no means a universal insistence on purely individually-based or non-racial representation. In his seminal essay on democracy in Africa, Sklar writes: '[It] would be mistaken to believe that the consociational idea of self-determination for self-regarding communities is counter-revolutionary *per se*. In so far as subnational group rights command general respect, democratic movements which disregard consociational precepts do so at their own peril'.[27] And Lijphart goes even further and defends constitutionally structured group participation, saying that *as long as agreement is reached* there is no reason why segmental voting registers should not work well.[28]

In short, there is no reason to state (as many liberals do) that the only way of choosing an accountable government is through open electoral competition. There has been a tendency, particularly among academics, to consider white communal representation as somehow obnoxious. However, this matter must be seen in a proper perspective. If whites, or any other group, consider themselves to be a communal group, the only real question revolves around whether an agreement can be reached with black representatives about a mutually acceptable system of representation.

People who insist on individually-based multi-racial representation would do well to take into account the perspective introduced by an eminent authority on nationalism, Elie Kedourie. Kedourie points out that, until recently, representation was understood to mean the representation of a group — a corporation, an estate or a 'nation' — which empowered its delegates to negotiate or act on its behalf. Kedourie concludes: 'It can be argued that this manner of representation is, so to speak, more representative of society and of the variety of interests which co-exist in any society, than representation on the basis of "one-person-one-vote"'.[29]

But attempts to establish a generally acceptable form of political representation will flounder if the sole result is to institutionalize Afrikaner and African nationalism (or a white and black communalism). Accordingly, those not aligned to any of the nationalist movements must make a concerted effort to build a third principle — one that can prevent the political arena from being dominated by the clash between two nationalisms. The third principle must involve a genuine attempt at nation-building and seek to combine the best talents of all South Africans, regardless of their ethnic background. This issue is taken up in the final chapter.

The abandonment of Westminster majoritarianism

Approaches to conflict-ridden divided societies have focused on ways of fashioning a new majority through the electoral process. It was hoped that the legitimacy generated by elections would impart authority to the new government, and that it would therefore be able to command respect and stabilize society. However, a divided society is divided precisely because of the dual sets of loyalty and accountability which its different communities have developed over time. After an exhaustive study of Northern Ireland, Padraig O'Malley penned this conclusion:

> [The] concept of the consent of a majority in Northern Ireland has no meaning, and formulae that invoke it, or

processes designed to bring it about, add another dimension to the problem because they raise false expectations, encourage the pursuit of outcomes that can't live up to their promise, and abet strategies that will only make a settlement more difficult.[30]

Kedourie uses similar terms when he points out that liberal democracies are wrong to consider their central tenets (the idea of one person one vote and the concept of rule based on the concept of the majority) to be universal, self-evident truths. In Kedourie's words:

> As regards the principle of majority decision, this is workable only on condition that majorities and minorities are variable, not permanent ... If majority and minority are perpetual, then government ceases to have a mediatory or remedial function, and becomes an instrument of perpetual oppression of the minority by the majority ... The worst effects of the tyranny of the majority are seen when ... the unalloyed Western model is introduced into countries divided by religion or language or race.[31]

Undifferentiated representative government assumes the existence of a nation-state, the legitimacy of which is not in doubt.[32] The fatal weakness of all suggestions for a liberal democracy in South Africa lies in the fact that, at this moment, such a nation-state does not exist and whites are unwilling to abandon their national identity and risk their security by entering into an open, competitive system.

It is unhelpful to regard this unwillingness as mere racism. An analyst of plural societies, Pierre van den Berghe, who is one of the strongest critics of apartheid, points out that majority rule can easily become a liberal veneer for racial domination. He concludes:

> If your constituency has the good fortune to contain a demographic majority, racism can easily be disguised as democracy. The ideological sleight of hand, of course, is that an ascriptive, racially-defined majority is a far cry from a majority made up of shifting coalitions of individuals on the basis of a commonality of beliefs and interests.[33]

In recent years, some German parliamentarians have also warned about the glib insistence that South Africa should move to majority rule. Otto Graf Lamsdorff of the Liberal Federal Democratic Party has argued that it is self-evident that white security forms the key to black freedom, while Egon Bahr of the Social Democratic Party has called for a unique model of coexistence with equal rights and special protection of minorities.[34]

The commitment of both sides to control their radicals

Any settlement would stand or fall on the mutual willingness of both sides to police the extremists within their own ranks. The heat of the conflict between extra-parliamentary forces and the ruling establishment has tended to bring to the fore personnel on both sides who are single-mindedly committed to the destruction of their opponent's capacity to function. We have seen organizations such as the National Education Crisis Committee which, at one stage, was prepared to negotiate with government, effectively destroyed by security action because a few of its members were considered to be militant activists and because of its morale-enhancing rhetorical posture.

We have also seen youthful activists in townships engage in forms of brutal confrontation which (however justified their feelings may have been) have obscured community mobilization around genuine local grievances.[35] While the ANC is engaged in what may be a serious attempt to formulate constitutional options worthy of consideration, even by the South African government, its military wing, through bomb attacks on civilian targets in the mid-1980s, has succeeded in whipping up a passionate resistance to the organization — even among fairly liberal whites.[36]

It is perhaps true that nothing focuses the political mind on conciliation as effectively as strong, unyielding and damaging opposition. In one sense, both township violence and strong security action have had their benefits. For conciliation to proceed, however, both sides in the conflict must be able to shift into a negotiation mode and signal this very clearly. This requires that both sides have control over the 'militants' within their ranks.

This requirement also implies that the ANC, which has become increasingly important in directing overall extra-parliamentary strategies, has to be legalized, possibly within a set of conditions which allow it full scope for constructive, solution-oriented mobilization within South Africa itself. While this may well weaken the powerful symbolism of redemptive 'liberation' with which the organization has become clothed, it would also begin to 'normalize' its image in the eyes of whites.

The organization may well run the risk of the fission of its ranks once it is able to organize legally within South Africa. On the other hand, the same may happen to the NP if it legalizes the ANC. If such steps allowed the conciliatory elements within these groups greater scope for the forging of constructive internal alliances, this would not necessarily be a bad thing. Above all, however, until both sides in the conflict realize, and find it possible, to reduce the role of the hawks and control the actions of the super-radicals, little possibility of a resolution exists — at least until such time as one or the other side is exhausted and in disarray. We do not expect this to happen soon.

Conclusion

The acceptance of these proposals for a transitional form of government requires a different approach to South African politics than that formulated by the internal opposition and external critics of the government. It would cast doubt on the political usefulness of unqualified majority rule in South Africa.

Reflecting recently on nearly two decades of 'troubles' in Northern Ireland, Bernard Crick, author of the definitive study *In Defence of Politics*, referred to earlier, wrote that people should realize that overall consent may not be forthcoming in such a society. He sums up the current consensus in Northern Ireland: 'everyone seems to agree that fundamental constitutional change in Northern Ireland would need not merely the explicit consent of a majority but also the consent of the minority'.[37]

The liberal democratic bias is so strong, however, that many observers refuse to accept this reality. Instead they threaten the Protestants in Northern Ireland and the whites in South Africa with the most dire consequences if they do not obey the wishes of the majority and the international community. Crick issues an apt admonition.

> There is a dangerous tendency, both in political and legal thinking, to move away from difficult but honest moral arguments into apocalyptic prophecies, masked as projections of social trends, of impending social breakdown if my case is not accepted entirely and at once. The debate ... is intense and deadly enough without such perpetual stage thunder.[38]

Only if these hard truths are accepted by all parties concerned will it be possible to consider a power-sharing agreement which meets with the support of both blacks and whites — and this is a crucially important requirement.

Notes

1   John Burton, 'The theory of conflict resolution', *Current Research on Peace and Violence*, no. 3, 1986, pp. 126 – 9. We deal separately with the issue of negotiations in Hermann Giliomee and Lawrence Schlemmer, (eds.), *Negotiating South Africa's Future*. Johannesburg: Southern, 1989, chs. 3, 13 – 14.

2   The best exposition of the South African debate is by the main modern proponent of the theory of consociationalism, Arend Lijphart, *Power-sharing in South Africa*. Berkeley: Institute of International Studies, University of California, 1985.

3   Patrick Chabal, (ed.), *Political Domination in Africa: Reflections on the Limits of Power*. Cambridge: Cambridge University Press, 1986.

4   Cited by Martin Staniland, 'Democracy and ethnocentrism', in *Political Domination in Africa*, p. 56.

5   Sklar, 'Democracy in Africa', in *Political Domination in Africa*, pp. 17 – 29.

6   Jean-Francois Bayart, 'Civil Society in Africa', in *Political Domination in Africa*, p. 110.

7   John Dunn, 'The politics of representation and good government in post-colonial Africa', in *Political Domination in Africa*, pp. 160 – 1.

8   John Lonsdale, 'Political accountability in African history', in *Political Domination in Africa*, p. 131.

9   For an analysis see Neville Alexander: *Sow the Wind : Contemporary Speeches*. Johannesburg: Skotaville, 1985, pp. 23 – 56, 126 – 53.

10 See the preface by Charles Carter to Desmond Rea, (ed.), *Political co-operation in divided societies: A series of papers relevant to the conflict in Northern Ireland*. Dublin: Gill and MacMillan, 1982.

11 This framework has been stimulated by Padraig O'Malley's effort to draw up one which could help to resolve the conflict in Northern Ireland. See his book *The Uncivil Wars: Ireland Today*. Belfast: The Blackstaff Press, 1983, see especially pp. 356 – 60.

12 *Cape Times*, 17 September 1987, p. 10.

13 *Cape Times*, 17 September 1987, p. 10.

14 Lonsdale, 'Political accountability in African history', p. 129.

15 Bernard Crick, *In Defence of Politics*. Harmondsworth: Penguin Books, 1982, pp. 18 – 21, 30 – 1.

16 Staniland, 'Democracy and ethnocentrism', p. 70.

17 Gerrit Viljoen, *Ideal en Werklikheid*. Cape Town: Tafelberg, 1978, p. 35.

18 Pallo Jordan, 'Why don't Afrikaners rely on democracy?' *Die Suid-Afrikaan*, no. 13, February 1988, pp. 24 – 5, 29.

19 For these distinctions, see Barry Buzan, *Peoples, States and Fear : The National Security Problem in International Relations*. London: Harvester Press, 1978, pp. 44 – 53.

20 Lawrence Schlemmer, 'South Africa's National Party government', in Peter Berger and Bobby Godsell, (eds.), *A Future South Africa : Visions, Strategies and Realities*. Cape Town: Human and Rousseau, 1988, p. 44. See also Helen Zille, 'The right wing in South African politics', in *A Future South Africa*, pp. 55 – 94.

21 No Sizwe (pseudonym), *One Azania, One Nation*. London: Zed Press, 1979, p. 101, citing *The African Communist*, Fourth Quarter, 1976, p. 115.

22 For a sensitive treatment of Inkatha see John Brewer, *After Soweto : An Unfinished Journey*. Oxford: Clarendon Press, 1986, pp. 338 – 406.

23 H. Puhe and K. P. Schöppner, *Views of Coalminers in South Africa on Sanctions*. Bonn: Deutsche Afrika Stiftung Publications Series, no. 45, 1987.

24 D. J. van Vuuren, et al., *South African Election 1987: Context, Process and*

Prospect. Pinetown: Owen Burgess, p. 235.

25   Study undertaken for Argus Group Newspapers by Marketing and Media Research, October 1988. Sample size: 382 Africans in the PWV area.

26   Oliver Tambo, Text of speech in Bonn, 1985, distributed by the ANC publicity department.

27   Sklar, 'Democracy in Africa', p. 25.

28   Lijphart, *Power-sharing in South Africa*, pp. 80 – 1.

29   Elie Kedourie, 'One-Man-One-Vote', *South Africa International*, vol. 18, no. 7, 1987, p. 1.

30   O'Malley, *The Uncivil Wars*, pp. 358 – 9.

31   Kedourie, 'One-Man-One-Vote', p. 3.

32   Chabal, 'Thinking about politics in Africa', p. 6.

33   Pierre van den Berghe, 'Introduction' to Pierre L. van den Berge, (ed.), *The Liberal Dilemma in South Africa.* New York: St Martins Press, 1979, p. 7.

34   Klaus van der Ropp, 'Nuwe benaderings nodig vir Westerse inisiatiewe', *Die Suid-Afrikaan*, no. 14, April/May 1988, pp. 34 – 6.

35   For a discussion of the role of the youth in the uprising see Mark Swilling, 'The extra-parliamentary movement: strategies and prospects', in *Negotiating South Africa's Future.* pp. 63 – 8, and Paul Zulu, 'Youth in extra-parliamentary opposition' in *Negotiating South Africa's Future*, pp. 69 – 74.

36   See *The Star*, 1 July 1988 for an example of the ANC's 'double-thrust'.

37   Bernard Crick, 'Northern Ireland and the concept of consent', in Carol Harlow, (ed.), *Public Law and Politics.* London: Sweet and Maxwell, 1986, p. 43.

38   Crick, 'Northern Ireland and the concept of consent', p. 42.

CHAPTER 8

The transition in operation : some proposals for consideration

In Chapter 7 we presented some of the major commitments, conditions and broad principles of compromise which we see as necessary for the resolution of South Africa's conflict. Since these are only ground rules, this chapter attempts to take the discussion closer to practical suggestions on steps towards reconciliation.

Right at the outset, however, we would like to make it clear that we do not want to propose yet another 'solution' for South Africa. This chapter makes concrete political and constitutional proposals simply to show that some kind of political accommodation is feasible. Our suggested scheme of alliance participation makes it possible for all the major parties and movements to participate in the process provided they step back from the demand to control the system exclusively and at the expense of their adversaries. Our scheme points to a third way: instead of domination by either Afrikaner or African nationalists, we propose a nation-building project. This has to be based on the genuine sharing of power and must have as its goal a transcendent South African nationhood with which all sections in the country can identify.

The idea of genuine power-sharing is not a purely academic construction but has in fact a broad range of support among South Africans across the colour line. A 1986 poll conducted among all groups in the highly politicized Pretoria-Witwatersrand-Vereeniging area (samples exceeded 1 000 per group) confirmed the results of previous studies and revealed a tendency toward conciliation. Respondents were asked what they thought life would be like under various types of government.

TABLE 8.1: PERCENTAGE OF URBAN SOUTH AFRICANS WHO ANTICIPATE
THAT LIFE WOULD BE 'GOOD' OR 'VERY GOOD' UNDER DIFFERENT FORMS
OF GOVERNMENT.
(Sample: 1 000 + per group in PWV area)

	Blacks	Coloureds	Indians	Whites
White-dominated government	13%	31%	54%	70%
Black majority government	46%	6%	5%	3%
Joint government without any group dominating	75%	68%	66%	58%

Source: D. J. van Vuuren et al. (eds). *South African Election 1987.* Pinetown:
Owen Burgess, 1987, p. 227; research by N. J. Rhoodie of the HSRC.

The results show that perceptions of either a white- or black-domi-
nated government vary markedly between blacks and the white/ col-
oured/ Indian minority groups. However, considerable appreciation for
the suggestion of joint government or power-sharing is evident. Not
one representative study along similar lines in the past has yielded
different results. Although representative popular attitudes do not
necessarily drive political action, these findings confirm that most
South Africans, apart from highly committed parties and lobbies, tend
to support compromise. It is also significant that the study was conducted at the height of the violent protests in early 1986.

To summarize the state of the political debate briefly: South Africa
is currently presented with two choices. On the one hand, the government envisages a gradual, tortuous, evolution towards an open society
in which power and wealth are shared increasingly more equitably than
they are at present.

On the other hand, the majority-based opposition movements, joined
by many academics and journalists, point to the near universal condem-
nation of apartheid and the contradictions of the system. They demand
that the government be moved from its party-political agenda, either
by force or moral persuasion. They display a great measure of confidence in the viability and future stability of an apartheid-free South
Africa. In the words of Heribert Adam: 'Shared languages, Christian
religious culture and consumerism ... should impart ... faith and optimism for the future.'[1] This assumes that power-politics and security
anxieties will not override these shared values between whites and
blacks.

Given the perceptions and fears in white politics and the nature of
politics the world over, we do not wish to make prescriptions on the
basis of faith and optimism such as that expressed by Adam and many

other authors. We believe that there is another and very powerful consideration which also has to be taken into account. Shared values and aspirations, as we have argued, do *not* negate powerful status and identity needs, which are often expressed at the expense of other groups in a divided society.

Whites, for various reasons, are a politically-mobilized and self-conscious group. For most of them, apart from the successful individualists in the upper-middle professional and intellectual classes, black majority rule would mean a disaggregation of their mobilized political 'corporate' identity. They would, in other words, suffer a loss of national or communal status. They would probably run very little real risk of victimization since their skills would make them a valuable resource (as is the case in Zimbabwe) with or without a bill of rights to protect their property and cultural interests. In many ways, however, whites would feel that they would have to defer to the symbols and political necessities of African majority politics. In symbolic terms, whites would become (privileged and perhaps protected) second-class citizens.

Recent research by co-author Schlemmer[2] has once again revealed the salience of communal identity. In a study of residential integration, even those whites who had no complaints about neighbours of another colour and whose property values had benefited from increased demand, felt uncomfortable in areas such as Mayfair in Johannesburg. In probing for completely spontaneous answers, it became clear that identity needs — 'soort soek soort' as the Afrikaans saying goes — overshadowed specific, material concerns in the way they approached day-to-day life in a shared society.

This does not mean for one moment that we reject the ideal of an open democracy or common society or, for that matter, the complete abolition of residential segregation. Our question simply is how to get there. Unless these needs and perceptions, however absurd or irrational they may be, can be accommodated, the majority of whites will not voluntarily move away from white domination. For this reason, we look for another route to change, basing our arguments precisely on what we argue to be a responsible and consequential *lack* of 'faith and optimism'.

The need for a 'buffer' period

We believe that to argue or plead for an abrupt and comprehensive transformation of the structure of political power in South Africa is fanciful, naïve or unwittingly revolutionary in its implications. Both

Zimbabwe and Namibia are, however, frequently taken as models (or emerging models) of precisely such a transformation. Presented in this way, these models are creating expectations of the possibility of a sudden shift.

What is conveniently overlooked in references to both the Rhodesia-Zimbabwe transition and the impending SWA-Namibia shift is the fact that both these territories had a very important 'buffer' phase in their recent history. Rhodesia, as we mentioned earlier, had the period of multi-racial government under Bishop Muzorewa and a constitutional acceptance by the dominant white party of an evolution to majority rule within ten years. Namibia has had a period of multi-racial rule under the Democratic Turnhalle Alliance, which has a mixed cabinet and has phased out statutory residential discrimination.

It is fashionable to discredit these buffer periods as 'co-option', 'too little too late' and so on. However, the acclimatizing experience of these transitional phases was crucial. Without them a 'transformation' free of massive popular resistance from whites would have been extremely unlikely. Muzorewa is commonly held up to ridicule, but we would argue that he was as important as Mugabe's war in facilitating the emergence of the new Zimbabwe.

It is understandable that politicians actively engaged in the struggle for change should attempt to discredit intermediate political processes for a host of strategic reasons, including the fear that the politics of the transitional period will become entrenched. It is, however, less understandable — indeed suggestive even of gross political irresponsibility — when supposedly experienced Western statesmen succumb to the strategic anxieties of liberation politicians and join the call for immediate 'transformation'.

It is precisely because a 'buffer' period of transition will facilitate the change towards an accountable political system in South Africa that we now proceed to suggest some of its possible components.

The basic requirements of transition

We have already outlined the broad principles of the kind of compromise we consider necessary in Chapter 7. More concretely, however, we would suggest that the following are the basic requirements:

- A government of national unity with a fully representative national assembly functioning on the principle that the support of both the majority and the minority has to be secured.
- The government has to allow the mobilization of 'progressive' non-racial parties. Unless it allows mobilization in opposition to

its own policies, its own legitimacy will forever be questioned.

- For the period of the transition, the white community and any other minorities which have the same needs should continue to elect their own representatives, preferably in some kind of minority alliance. Individuals within a minority group should, nevertheless, be free to support a 'progressive' non-racial party.

- The guiding political principle should be one of voluntary association and not formal racial-group definition. Political parties or alliances should be the main vehicle for the articulation of group interests. We believe that the following party formations or alliances cover virtually the entire political spectrum in South Africa:

  — *a majority alliance*: this will stress non-racial, individual rights and redistribution of wealth in a system of social democracy;

  — *a minority alliance*: this will advance the political and cultural claims of minority groups and will defend the interests of property owners, business and many of those employed by the existing state;

  — *non-aligned formations*: these will be comprised of groupings such as workerists/Marxists, who propose control by workers; liberals who may prefer a classic Westminster winner-takes-all system, and nationalists who clamour for an exclusive African (or white) homeland.

- The strongest white party (whatever it may be when transition occurs) should articulate white interests in the minority alliance. It need not, indeed, should not be exclusively white. The evidence that we presented earlier suggests that it will not be.

- The principle of sectional domination, either by a majority or a minority, should be curtailed and, along with it, any institutionalized factional or group advantages or privileges.

- The underlying conflict is increasingly not between white and black but between minority and majority interests and needs, and should therefore be resolved in such a way that the support of both the minority and the majority is secured.

The absolute prerequisite for the proposed scheme is for the NP government to abandon its insistence on basing the entire political system on racially defined voters' rolls. Archbishop Desmond Tutu, amongst others, is obviously correct in stating that the main bulwark of apartheid is race classification according to the Population Registration Act. There can be no moving away from discrimination based on race unless the law which states that people will be classified, whether

they like it or not, according to race, is removed from the statute books.[3]
We do not accept the NP's argument that race classification is necessary
for the stability of South Africa. Once alliances are in place the
abolition of race classification can occur, and a non-racial system of
government can replace apartheid.

The most obvious way in which South Africa could enter into a
transitional system would be along the following lines:

- The party enjoying majority support among whites should open its
  membership to people from all groups and seek common cause
  with all other parties and organizations that articulate minority
  concerns. The goal should be a *multi-party alliance or pact.*
  There is, as we have noted, a precedent for this in Namibia in the
  Democratic Turnhalle Alliance which, even though it may end up
  in opposition, seems set to remain a very viable political formation
  in that territory.

- The alliance should then invite the other 'tradition' (the individualistic, majority-based formations) to agree in principle to enter
  a government of national unity, based on the principle of parity of
  representation in government between the two formations (a 'majority' and a 'minority' alliance) for a transitional period of at least
  ten years. At the end of this period a constitutional commission
  would review the arrangement and prepare the way for regular
  elections within an appropriate framework.

- We recognize that the dual structure of transition we propose will
  create problems for minority-based parties which have as *part* of
  their programmes the articulation of majority interests. For instance, where would the Democratic Party or the Labour Party
  choose to sit — with the majority or the minority side? The
  position of Inkatha would be equally complex. We would suggest
  that *if* these parties find it inappropriate to choose sides, as it were,
  then the opportunity should be created for them to participate from
  an independent position. Representatives of non-aligned parties
  could be chosen with a minimum poll to eliminate fringe parties
  or candidates. Either as a collective, as an alliance, or as single
  parties, they would form the opposition in the government of
  national unity.

- For the purposes of participation in the transitional government of
  national unity, the 'alliances' might organize separate elections to
  choose their representatives for a national assembly. These elections of representatives would be similar to the party primaries in
  the United States, in the sense that people would have a choice to

vote for representatives in either the majority alliance, the minority alliance or the non-aligned (mediating or opposition) parties.

- Parliamentary representatives of non-aligned parties should not be allowed a casting vote but should play an oppositional or mediating role. Should the non-aligned parties draw a larger number of votes in a poll than either of our conceived major alliances, then obviously one or more of these non-aligned parties should get a proportional share in the government of national unity.

- The national assembly or parliament would be based on a simple equal division between the majority and minority alliance subject to the qualification above. Neither would be (or should be) precisely defined. Obviously, however, the white majority party would tend to be the largest formation within the minority alliance and the ANC the largest grouping within the other.

- Ground rules for political mobilization would have to be formulated by a panel of judges and other experts, and would have to include an amnesty for the ANC and the release of most categories of political detainees and prisoners. The Western nations could play a valuable role if they were to accept the ground rules, and then use their influence and ability to keep a watching brief on the SA government, the ANC and other liberation organizations during the period of mobilization and implementation. The latter should be invited to nominate representatives to the panel of lawyers formulating the ground rules.

- The emphasis during the transitional phase should be on *accountability* and not on demographically-based politics of numbers. Representatives should, in the first place, be accountable to their specific alliance, but as a new power-sharing government takes root, it should establish ways of developing more general accountability. Referendums on major issues would be one possibility.

- South Africa is a complex society with a great deal of regional diversity, and our proposals for a transitional government of national unity must accommodate regional influences. For this reason it would be necessary to establish a second chamber, constituted much as the United States Senate is structured, in which the two or three parties gaining most votes in particular regions would sit. The demarcation of regions, which should include non-independent homelands and, if agreed to by all, independent homelands, should be based on negotiation between the alliances in the assembly. The second chamber would therefore be formed after the establishment of the national assembly. Its powers should follow the pattern of the United States Senate, which has proved itself admirably in a complex political system. The specific benefit and task of this senate would be the re-unification of South Africa's fragmented political geography.

The kind of system we suggest should last as long as required to allow both the fears of minorities and the heightened expectations of the majority to subside. We suggest ten years. After this period it is quite conceivable that realignments and new political divisions would have occurred which would enable the system to move towards a more normal democracy. In any event there are various constitutional models which could be assessed at the end of a buffer period; models which are capable of accommodating communal aspirations. As we have suggested, a constitutional commission would assess and recommend whether the transitional period should be prolonged or phased out and what system should replace it. A decision in this regard would have to be based on general agreement.

Even within this kind of transitional arrangement, political realism requires that the ambitions of all the leaders be channelled and checked by more than principles and initial commitments. Because of the large material inequalities in the country, and the extent of black aspirations and white fears, the political system will put all leaders under great pressure to protect established interests or, conversely, to demand the dismantling of any structure which does not meet with the approval of their followers. The stresses of socio-economic change, together with the aspirations these engender, will constantly make immediate, populist claims on all leaders, whether minority or majority based.

For these reasons a resolution based on the kind of principles and compromises we suggest, can probably not be achieved without the introduction of certain policies calculated to reduce the social and economic strains for which South Africa has become so well-known. We turn to a brief review of these issues.

Intermediate policies to support transition

Among the many strains which have to be ameliorated before a solution becomes workable, we can identify the following major issues.

The elimination of the remains of segregation

Residential segregation is perhaps the most pervasive and visible aspect of apartheid. The effect of apartheid and segregation has been that the most favourably located, well-established and best-serviced residential areas are those of whites. The perpetuation of this form of inequality will undermine any agreements between political leaders and prejudice compromise and co-operation. The current conflict in the Tricameral Parliament is ample proof of this.

.Unfortunately, there are no easy solutions. A complete abolition of all racial zoning of residential areas is unlikely to have much more effect than the civil rights campaign on residential patterns in the USA. There is an abundance of incontrovertible evidence from that country that residential segregation between black and white Americans has strengthened rather than declined over the past twenty years.[4]

On the basis of a detailed examination of residential trends, co-author Schlemmer[5] has concluded that a carefully managed 'dispersal' of black settlement in white areas is likely to produce the required variety and quantity of housing opportunity that blacks need and desire. If existing residents are not given the assurance that the social character of their neighbourhoods will not change rapidly, mixed residential areas will simply evolve into black ghettos while new and near exclusive white areas will spring up. The phemenon of 'tipping', in terms of which whites flee a neighbourhood once the proportion of new (black) residents reaches roughly 15 to 20 per cent, depressing property values and lowering standards, is as likely to occur in South Africa as it has in the USA or Europe.

The extreme frustration which blacks with aspirations for better housing and a well-ordered suburban environment are likely to experience if new, informal but pervasive segregation emerges, will quite clearly radicalize their political sentiments.

For these reasons we suggest that, for a period of transition, desegregation should include active measures aimed at the maintenance of the established social (not racial) character of neighbourhoods and local participation in decision making. This should occur after as much agreement as possible about procedures and standards has been reached in negotiations between the government and other major political movements within South Africa. A simple, categoric demand that Group Areas be abolished forthwith will not achieve much over the short to medium term. It is for this reason that we suggest a managed transition, which stands a reasonable chance of winning the support of whites rather than their intractable opposition.

Much the same can be said of the desegregation of schooling, which obviously goes hand-in-hand with residential desegregation. If the process is linked to the tempo of managed residential desegregation, it might be possible to avoid the rapid transformation of some schools from all white to nearly all black — a tendency established decades ago in the USA.

The expansion of urban opportunity

One of the functions of rigid apartheid under Verwoerd was to keep the Third World at bay and, in so doing, protect the essentially European character of South Africa's cities and towns. The economy adjusted to this by means of the migrant labour system which accommodated economic interdependence without allowing this to impact drastically on the formal socio-cultural setting of the major towns and cities.

As we have described in Part One, the 'fences' between the First and the Third Worlds, or between South Africa's two 'colonial' segments, began breaking down in the late 1970s under the pressure of mounting unemployment in the homelands. By late 1987, it was estimated that there were over two million black squatters around the major metropolitan areas of the PWV, Durban, Cape Town and Port Elizabeth.[6] The resistance of the squatter communities and the turmoil of reaction and counter-reaction by 'vigilantes' and comrades which followed, particularly in Cape Town, plunged the issue of the marginal urban-dwellers (South Africa's Third World urban fringe) into the international spotlight. The 'peripheralization' of South Africa's poor failed as a strategy, like so many other aspects of separate development. The abolition of influx control in 1986 was a formal admission of this failure.

Recent research done by co-author Schlemmer and others, reveals the full implications of the new urban form which is emerging. In the Pretoria-Witwatersrand-Vereeniging area, for every 100 formal black township houses, there were 70 'backyard' shacks and 20 garages used as additional accommodation for people who could not find or could not afford formal housing.[7]

The government has taken steps, rather belatedly, to identify land for site-and-service housing, and in terms of quantity, the short-term needs are probably gradually being addressed. Two major problems are likely to persist.

First, local leaders testify that there is great sensitivity among surrounding white and Indian residents about the proximity of black squatters, hence the land areas being identified are futher and further from areas of employment. For example, one new area called Orange Farm, which is due to absorb squatters from a variety of areas in and around Johannesburg, is closer to Vereeniging than to Johannesburg.

A second problem is that the service charges on the vacant stands with rudimentary services are beginning to approach the full rentals and service charges for older, subsidized township houses and, in some cases, exceed the municipal rates payable in white areas. Furthermore, one may confidently predict that as the existing squatters are catered for, many more people will move into the cities, not only from the homelands, but also probably from Mozambique and Zimbabwe. Already there are small communities made up entirely of Mozambicans near Johannesburg.

This brief outline of the problem indicates another serious source of discontent and potential instability which will threaten compromise between existing white territorial interests and the demands of black leaders entering the system. The problem remains one of dealing with the interface between informal Third World communities, which are destined to become a majority in the cities, and the mainly middle-class white, coloured and Indian communities who are the established insiders.

Close juxtaposition of these two kinds of urban settlement patterns is unlikely to be accepted by whites. One can see right-wing groups making considerable political capital out of the friction, thereby adding to the burdens which the politics of compromise has to carry.

What is required, apart from a substantial increase in available land, is housing for the poor. The peri-urban squatting concentrations have to become fully absorbed into a framework of administration which is sensitive to the particular socio-economic needs of these communities. It goes without saying that unemployment and under-employment is higher in these communities than it is in established black townships. One needs regional authorities considerably more developmentally oriented than the current Regional Services Councils. The squatter belts have to become the target for state-facilitated industrial development strategies which emphasize labour-intensive production. The communities also have to be employed in the development of their own bulk-services, social services and recreational amenities.

Needless to say, the problem of boundary friction between these communities and adjacent formal areas can only be dealt with if a firm but legitimate community authority is established, requiring full participation by the communities themselves. There is no space left in the South African cities or peri-urban areas for large so-called buffer-zones, which used to be the solution in the past. These communities have to police themselves in order to control the extent of the petty-crime and trespassing which is the core of the present problems on the boundaries.

We are, therefore, suggesting that a new and comprehensive form of urban and peri-urban development corporation, be established as a division of or adjunct to the Development Bank of South Africa, to ensure the successful upgrading and legitimate social control of South Africa's urban Third World. Just as regional development corporations were established in the homelands to facilitate their political development, so an urban development corporation should assist in easing South Africa's transition away from apartheid.

Economic development: the need for a statement of political intent

The upgrading of urban opportunity for blacks will simply not be possible without more dynamic economic growth within the country. At present our growth is subject to a low ceiling caused by balance of payment constraints. One issue at the heart of the sluggishness of the economy is a lack of foreign capital, partly caused by the sanctions campaign.

This campaign will only be lifted if the government is able to give the outside world a persuasive statement of intent regarding the political incorporation of blacks. The statement of intent can be linked to the promotion of the need for development capital, specifically for use in upgrading and expanding black urban opportunity.

A programme along these lines, coupled with a special black development fund, will probably not generate much enthusiasm abroad unless it is managed and controlled by an institution other than the government itself. We suggest that a body comprised of government representatives, private sector organizations, representatives of black communities (formal and extra-parliamentary) as well as nominees of certain foreign governments, might be established to manage the programme.

South Africa itself has to address a further major impediment to reconciliation. This is the inequality in ownership of capital: at present more than 90 per cent of capital is in white hands. However, conventional methods of redistribution by a central state authority would curb investor confidence and reduce growth. The majority would experience only a short-term benefit.

Two approaches suggest themselves. One has already been launched by certain companies. This is the issue of shares to employees. This could occur either through the involvement of trade unions or the direct participation of individual employees, depending on the dynamics of industrial relations in different companies and sectors. Consideration might even be given to granting tax concessions to companies in order to encourage this type of investment in future stability.

A second, perhaps complementary approach could be through the urban development corporation suggested earlier, which would receive a basic issue of share capital from government, and which could buy into major corporate interests, and then sell share-equity to blacks at favourable rates.

Intermediate political steps

A statement of intent, as we have outlined, will not carry weight unless there is a visible and immediate shift in political structure, which would provide a powerful symbolic demonstration of a movement towards resolution. Obviously an interim shift is not likely to encompass the broad elements necessary for an overall resolution. Two developments, however, are in our view both possible in the near future and very necessary.

One concerns the present constitutional device of 'own' and 'general' affairs. The emerging reality of integrated residential areas, together with the need for a concerted, integrated housing strategy for the urban areas as a whole, and the gathering momentum of private open schooling, are all steadily undermining the functions implicit in the concept of 'own affairs'.

We imagine that the government will not summarily dispense with 'own affairs' as this is its only device with which to counter accusations from the right-wing that white interests are being sold out. What is necessary, however, is for the category of 'general affairs' to be steadily augmented so as to encompass those institutions under the purview of 'own affairs' which are subject to processes of integration, such as municipal and local authority affairs, housing, common amenities and services, and an expanding core of open education. 'Own affairs' can remain as the constitutional 'stream' dealing with services and amenities, including education, for communities which deliberately choose not to become part of the transition which is currently occurring. Conceivably, 'own affairs' functions could, at the level of administration at any rate, be handed over to the various communities to run. In this way a programme of devolution could avoid sharp conflicts over the deployment of central state resources.

The second requirement is that where integrative processes are clearly advanced, and where a majority of each population group in the situation is willing to accept the process, the transition should be taken to its logical conclusion as speedily as possible. The present trend is for evolutionary processes to occur with a great deal of vacillation and ambiguous rhetoric on the part of government, which both raises and frustrates black expectations. There is a great need for certain major reforms to occur with sufficient expedition to 'disarm' the critical attitudes among leaders in the black communities.

The benefits of an acceptance by government of the KwaZulu-Natal Indaba, albeit with negotiated modifications, goes virtually without saying. As important, however, would be to mirror this indaba process in certain cities and towns where the political climate would be supportive of the full implications of re-uniting the divided urban structure. Cities which spring to mind immediately are Cape Town, East London, Durban, Grahamstown and Pietermaritzburg. Others would inevitably follow. In such areas this would imply the expansion of the local authority system to incorporate all groups (with safeguards if necessary) into a united system of local government.

This could be a logical and constitutional extension of the government concept of 'Free Settlement Areas' to embrace entire urban areas. We believe that this would make an impact on audiences abroad and make the internal black leadership more inclined to co-operate in a process of change.

Towards a dual nationhood

Before turning to our final comments on nation-building as a prerequisite for an inclusive democracy, a few words on a much-debated alternative route are required. We refer here to the principle of devolution of powers advanced, as we have seen, in the federal policy of the PFP, and more recently persuasively expounded by Leon Louw and Frances Kendall in their work *South Africa: The Solution*.

Very broadly, the argument is that the conflict of political goals (likely to take place at the centre in a unified political dispensation) would be reduced in scope, fragmented, or made subject to counter-balancing regional variations if legislative and executive authority were to be devolved downwards to smaller geographic units. It is also argued that government can be brought closer to the people, indeed, can be made subject to Swiss-style popular referenda; that local political groupings are more likely to recognize their immediate interdependence with other groups under such a system; and that popular interests at the local level can constitute a vital market-place of policy options which will erode mass-based national ideologies and hence the basis of the major conflict in South Africa.

We would not disagree with most of the benefits claimed for

devolved systems of government, and endorse the general argument
that federalism or 'cantonization' of government improves the prospects for a democratic order which is responsive to community needs
and interests. We, therefore, have no difficulty in supporting the ideals
of devolved political authority. Our neglect of this course of development is simply due to the fact that the political struggle in South Africa
is dominantly over the nature and composition of *central* state power
and there seems almost no prospect of exploring decentralized or
devolved government until the conflict surrounding national government has been resolved. Although devolution should be on the agendas
of the main contenders in the South African conflict, it is not.

The present government has a fairly long-standing commitment to
decentralization of authority and functions, but the initiatives embarked
upon thus far (such as the Regional Services Councils, or the Regional
Development Advisory Councils) are either very limited in scope or
vigilantly supervised by the central executive or centrally-nominated
provincial authorities. Indeed, the high stakes in political policy in
South Africa induce an ultra-caution in government which, almost by
definition, emasculates decentralized structures. Simultaneously,
there has been a centralization of some sensitive political functions, as
seen, for example, in the transfer of white, coloured and Indian education from the provinces to the Tricameral Parliament.

On the other side of the conflict, in ANC political thinking, there
appears to be no rhetorical gesture in favour of devolution. The goal
of achieving a non-racial and *unitary* system appears to be unshakable.
Indeed, the kind of restructuring that the ANC would consider itself
obliged to attempt to implement would require centralized executive
authority, *par exellence,* and would, by all accounts, entail abandoning
the decentralized structures, of the homelands at least.

Regrettably, our arguments towards a resolution of conflict in South
Africa have had to be based on the possibilities inherent in the conflict
as it has existed for decades and as it is likely to persist for some time.
There are most definitely more fortunate constitutional models in
existence. We claim no particular constitutional elegance, efficiency
or theoretical appeal for our proposals. Their validity lies very crudely
in the fact that they attempt to take the parameters of the conflict as it
exists in our day-to-day political reality as a starting point.

The devolution of power to smaller geographic units is also not seen
by the National Party as being able to address the key issue of white
minority rights and the integrity of white political institutions. The
simple reason for this is that there are only six magisterial districts in

the country in which whites are a majority. It is this factor which contributed to the government's extreme reluctance to allow a federal precedent in the form of the KwaZulu-Natal Indaba proposals. While we are pessimistic about the chances of achieving effective devolution before the major power conflict at central government level is resolved, we nonetheless would suggest that regional negotiation initiatives may be part of a very important process towards the resolution of central power. If one assumes that the government, the ANC, Inkatha and other political players are in a very early pre-negotiation stance, they will all be reluctant to make concessions or commitments in terms of their 'central' policy positions which might weaken their bargaining power later.

It is, however, very necessary to demonstrate the viability of mutual accommodation at some level which will not be seen to commit the parties irrevocably to the type of compromises which may be involved. Negotiations at local level or regional level can provide such a demonstration (and may indeed further the prospects for devolution as well). The KwaZulu-Natal Indaba is the classic example, although it has yet to draw in the extra-parliamentary movements, but others are required. It is particularly important that UDF-ANC organizations be involved. Some areas of the country which are somewhat removed from the critical tensions of central conflict-management, present themselves as possible favourable locations. The Eastern Cape could be an area where a new regional initiative could be usefully pursued. With these comments we return to our final observations.

We are aware that there will be critics who will immediately reject our proposals as being, at best, hopelessly inadequate for national reconciliation and, at worst, a strategem for continued white domination. Others may not be as hostile to our proposals, but may argue that they cannot work because one would need a common conception of nationhood, along with shared political values in order to make them work.

We agree that there is, at present, no common nation in South Africa. Since our suggested minority alliance will be dominated by whites and the majority alliance by Africans, and since the huge economic disparaties between whites and Africans cannot be bridged overnight, there is a danger that the two alliances will become so polarized that political bargaining is ruled out. If the process of reconciliation were to fail, the society might emerge even more deeply divided than ever before.

We are, nevertheless, convinced that it will be best for the process of reconciliation if the ethnic, communal and nationhood issues are openly faced and a genuine attempt is made to accommodate them. It is in fact only when a group has the assurance that its political and national identity is guaranteed that the question of a division of power becomes negotiable.[8] Comparative studies on societies such as India, Malaysia, Lebanon, Ghana and Nigeria show that it is precisely the refusal to openly recognize the communal and national realities that gave rise to major turmoil and bloodshed. This happened, for instance, in India where the Gandhi-Nehru tradition considered communalism as a pathological condition of the body politic.[9]

For these reasons it is important to address the vague but nevertheless powerful symbolic issues which cluster around the national question. For example, Afrikaner or more general white interests, at a stage when negotiations around power are imminent, will probably turn out to have both economic and status components (fears of occupational discrimination and violation of property rights, for instance) as well as cultural or symbolic components. The former fears could be assuaged by, for example, a bill of rights or anti-discrimination clauses; in other words, a negotiated resolution will be possible.

The same process of resolution may very well not be effective with regard to the zero-sum symbolic issues. Will whites or Afrikaners be prepared for national 'demotion' from the status of a leading group and a power establishment? Will they easily exchange a historical symbolic heritage of identification with the South African nation-state for the status of a mere minority or class category? Many symbols would be at risk in the process, from flags and place-names, to thousands of small towns and suburbs with an entrenched ethnic character, extending to even more sensitive issues such as culturally-infused education.

Needless to say, the substantive content of these issues is usually flimsy or mythical but, prior to the 'sacrifice', they are perceived as essential collective *needs*, rather than as quantifiable and negotiable interests.

On the side of the African majority there are equally compelling national symbolic needs. It is undoubtedly true to say that in most African societies to the north, rank-and-file (as opposed to élite) material interests have suffered since independence. Yet, it is doubtful that any popular poll would reveal a collective desire to reverse the event of 'liberation'. Team fervour, national honour and the opportunity to identify, no matter how remotely, with the leading power establishment is the poor person's equivalent of the material and status rewards of power.

The African majority in South Africa will without doubt wish, indeed need, to see the society changed from the country of European or white South African imperialism to a country which will become a liberated nation of and for Africans. The population may consist of both black and white, but the popular expectations will be that everyone will have to defer to African symbols. The need for this symbolic reversal will be just as powerful as any equivalent need for the maintenance of symbols and national honour among whites.

It is this aspect of the South African conflict which requires resolution in a larger context than that allowed by a piecemeal approach — which, at best, would be a long drawn out and tedious process. Indeed, it may even be argued that it would be desirable to address symbolic issues so that the tensions caused by both positive and negative expectations are lowered prior to any concrete negotations. If this does not occur, the vague and imprecise but powerful symbolic needs may 'pollute' or even usurp interactions which might otherwise be constructively addressed.

We suggest a set of dual national symbols to represent the two historical streams in South African political life and as a gesture of national accommodation. In deeply divided societies, such national accommodation becomes possible only if the legitimacy of communal and/ or sub-national attachments is accepted and these are seen as positive second-level loyalties which contribute to a national unity based on diversity. In this context, agreement must be sought about a dual set of national symbols which would symbolize the unity in diversity. South Africa has a historical precedent for this. In the period 1910 to 1960, two anthems (*Die Stem* and *God Save the King*) and two flags helped Afrikaners and the English-speaking community to come together under the Union of South Africa.

There is the same need for ideological dualism. We believe the basic issue is the common acceptance of the main traditions and principles which have taken root in the course of South Africa's political history. One of the ways in which this can be achieved is if both sets of protagonists begin to modify their basic political charters in order to allow political space for one another in a future dispensation. In other words, without necessarily sacrificing their own charters, they might consider allowing space for the alternative charter.

As we have argued, the most fundamental commitment of the resistance and black opposition movements, and one which they can only sacrifice at the cost of their political essence, is the principle of *non-racialism*. This is the principle of an open society in which the divisions of race and ethnicity, which have relegated blacks to a real and perceived second-class status, are swept away. It is also the concept of an open society which, given the numerical preponderance of blacks, will unlock the symbolic ideal of a liberated African state.

As already suggested, the most fundamental commitment of the great majority of whites is to *participation as a communal group* in the political process so as to retain control over their own material, social and political destiny. Having been politically mobilized as a group for centuries, they fear being ruled, however benignly. Even allowing for the fact that almost all white fears of majority rule are based on myth, political retirement, like occupational retirement, is traumatic to contemplate. The results of a poll referred to earlier (Market and Opinion Surveys, November 1986, nation-wide sample 1804) show that no more than 3 per cent of Afrikaners and 8 per cent of English-speaking whites were prepared to countenance an unqualified majority rule dispensation in South Africa.*

National dualism is not to be equated with institutional ethnicity, as exemplified by present policies or even with more flexible forms of consociationalism. One of its two elements must be a genuinely open sphere of political and social life and this ultimately will have to be the dominant sphere in South Africa. Whether this non-racialism be accommodated in major regions, or at particular levels of government, is not the immediate issue; the issue is that it is a universal value which whites would do well to accept as both necessary and inevitable for the larger South Africa.

The other element, communal self-expression *based on voluntary association*, is internationally accepted for Cypriots, Sri Lankans, French Canadians, Belgians, Israelis, Soviet Armenians and many other sub-national groups, too numerous to mention. The resistance movements need not fight this principle of political organization if it allows major 'space' for the non-racial option. In other words, even in the period after transition one should conceive of communities based on voluntary association, which enjoy sufficient devolved authority to maintain a communal existence.

A process of negotiation can set South Africa on the road to an accommodation between whites and blacks. If, however, one of the parties simply uses the increased political space that will necessarily

---

* Although it may be true to say that polls cannot forecast actual choices if the leadership becomes committed to persuading followers to change direction, they do, however, reflect the scope for an active mobilization by whites against sweeping concessions, unless the powerful symbolic needs are taken into account.

be created by negotiations, to demonstate its nationalist commitment and impose its hegemony,[10] the negotiating process will collapse. An aspect often overlooked in the discussion of negotiations is the fact that there must be a fundamental political settlement of basic issues before real progress can be made about specific issues. Furthermore, negotiations can only be successful if each side is prepared to give up something in order to gain some other objective.[11] In the South African case, this means a mutual acceptance of national dualism and ideological pluralism. Unless political leaders of all the main parties and all the movements accept this, it would be hopeless to attempt to implement a scheme such as the one we have proposed in this chapter — a transition to a government of national unity.

Conclusion

Towards the end of his life, Albert Camus, a native of Algeria, was under strong pressure from the French Left to throw his great moral authority behind the bitter struggle of the Algerian nationalists. He wrote:

> 'You must take sides', cry out those stuffed with hatred. Ah, but I have taken it. I have chosen my own country, the Algeria of the future where French and Arabs will associate freely together.[12]

In South Africa there is a similar tendency among spokespeople for both government and the liberation organizations to insist that people take sides. Behind this there is an unspoken demand to choose between Afrikaner (or general white) nationalism and majority-based nationalism as articulated by the ANC. Such demands posit a false dichotomy. The real choice is between the accommodation of the main streams in South African political life and unending strife; it is between building a transcendent South African nation and endemic conflict between Afrikaner and African nationalism. In Camus's terms, it is possible to build a future South Africa in which whites and blacks will associate freely. It is up to the people of South Africa to realize this vision.

Notes

1   Heribert Adam, 'Towards a democratic transformation in South Africa,' Cape Town: Institute for the Study of Public Policy, UCT, January 1989.

2   L. Schlemmer and L. Stack, *'Black, White and Shades of Grey'*. Johannesburg: Centre for Policy Studies, 1989.

3   *Perspective* (University of Cape Town), Interview with Archbishop Tutu, March 1989, p. 8.

4   For a review of US literature see Johan Fick et al., *Ethnicity and residential patterning in a divided society: a case study of Mayfair*. Johannesburg: Rand Afrikaans University, 1988.
5   Schlemmer, *Racial Zoning: Problems of Policy Change*. Johannesburg: Centre for Policy Studies, May 1989.
6   See *Sunday Star*, 7 August 1988, for a report by the Urban Foundation on the extent of squatting.
7   L. Schlemmer, L. Stack and M. Wholley, 'Urbanisation and squatting on the PWV' (Provisional title). Jhb: Centre for Policy Studies, (forthcoming).
8   Theo Hanf, et al., *South Africa : Peaceful Change*? London: Rex Collings, 1981, p. 380.
9   Ratna Naidu, *The Communal Edge to Plural Societies: India and Malaysia*. Ghaziabad: Vikas Publishing House, 1981. See also David R. Smock and Audrey C. Smock, *The Politics of Pluralism : A Comparative Study of Lebanon and Ghana*. New York: Elsevier, 1975, pp. 304 – 34.
10 This point is well made in Benjamin Nimer, 'National liberation and the conflicting terms of discourse in South Africa : an interpretation'. *Political Communication and Persuasion*, vol. 3, no. 4, 1986, pp. 313 – 53.
11 Daniel Katz, 'Nationalism and strategies of international conflict resolution', in Herbert C. Kelman, (ed.), *International Behaviour : A Social-Psychological Analysis*. New York: Holt, Rinehart and Winston, 1965, pp. 382 – 85.
12 In Philip Thody, *Albert Camus*. London: Hamish Hamilton, 1961, p. 214.

Postscript: February 1990

This book was first published in August 1989. Since that time political observers throughout the world have been astounded by the precipitous collapse of East European socialist governments. Equally striking has been the explosion of ethnic minority feeling both in Eastern Europe and the Soviet Union. We have witnessed a passionate determination among language, religious and national minorities to assert their political autonomy and to defend group boundaries, often with little regard for the economic implications of their campaigns.

In the light of these events it is ironic that some of the sharpest criticism of our analysis in Part 2 has come from scholars who have maintained that we have overrated white minority ethnic interests and political identity needs. Heribert Adam,[1] for example, rejects the weight we give to ethnic minority agendas, arguing that 'it is highly doubtful that the majority of whites would want to defend their position as if they were a national identity under threat'. He is convinced that, *inter alia*, the 'lure of the pocketbook' will substantially mollify the ethnic interests and reduce the political priorities of whites. In the ranks of the liberal white opposition, the Democratic Party (DP), he discerns 'unequivocal commitment to non-racialism' and a 'decisive lack of concern with identity'.

If Adam and other critics were right, one wonders what the decades of conflict in modern South Africa have been all about. The optimistic claims of these critics conflict with the available empirical evidence. As far as we are aware, no single representative political opinion poll conducted among whites has shown that more than 10 to 15 per cent of whites are prepared to entertain an undifferentiated mass-based political system. In a nation-wide poll conducted in 1989 among a representative panel of 1 800 whites,[2] even among DP supporters only 5 per cent endorsed the concept of 'one-person-one-vote in a unitary state'. These Democratic supporters, while disliking the idea of an overtly race-based system, signified their fears of becoming politically eclipsed in a mass democracy by opting overwhelmingly for a federal system, which would be negotiated to protect the importance of minorities as established political constituencies. The overwhelming majority of whites persistently appear to demand the right to maintain political coherence in some form or another and to choose their own leaders.

This book takes these realities fully into account. Its conclusions are based on the premise that without the committed and willing participation of not only individual whites who are active in the change process, but also of whites as a definable and mobilized constituency, an alternative to the present system of white domination will not be workable.

We are in no sense arguing that some form of immutable, visceral or primordial identity drives white interests. It is simply a universal fact that if a category of persons has social interests to defend and has become mobilized within social boundaries, that category of persons will attempt to maintain those interests and boundaries with determination. It should be noted that whites have been mobilized as a separate political entity at least since South Africa became a Union.

Nor are we arguing, as some critics have suggested, that whites have valid reasons to believe that a majority-based government will be fundamentally undemocratic. No one really knows whether or not a future government, faced with enormous socio-economic problems and possibly with regional or ethnic dissent, will be able to remain committed to democracy. The relevant political fact is simply that there is no way in which it can be 'proved' to minorities that a majority-based government will remain democratic.

However, this book was written in 1989. We must assess whether or not more recent events should qualify our analysis.

The advent of a new white leader in the person of Mr F. W. de Klerk has significantly changed perceptions of both the timing and content of the future compromises with mass demands which the South African government can be expected to reach. The release of Mr Nelson Mandela, the world's best-known political prisoner, the unbanning of mass-based resistance movements like the African National Congress (ANC), the partial lifting of the State of Emergency, and the official acknowledgement of the right to public protest have given rise to widespread expectations that white minority power is on a countdown to extinction. This expectation has been intensified by the sudden power eclipse among Eastern Europe's exhausted and politically played-out Communist parties.

This new context has fuelled the benign, indeed admirable, sentiment that a form of unqualified, mass-based democracy is at last achievable in South Africa in the short to medium term.

Certainly, and as we suggest later in point 4, superficial, incorrect interpretations of what the spokespersons of government have been saying lends some credence to this view. In his speech at the opening of Parliament in February 1990, De Klerk, for the first time, eschewed the use of the term 'group rights', and used 'minority rights' instead. Both he and Foreign Minister Pik Botha have pointedly emphasized that the new constitution they wish to negotiate will have to be approved by an overall majority of the population, without making any reference to ethnic subdivisions in the country.

Both our personal and political commitments and our perceptions of what would be most appropriate for South Africa in the longer term incline us to endorse the non-racial or unitary ideal. Indeed, we would argue strongly that, after a period of normalization in our politics during which the currently latent variety of political interests in the black communities can emerge, South Africa will have the balance of cross-cutting divisions and commonalities to make an individually-based unitary democracy viable. This consideration, however, does not address the *short- to medium-term transition*.

There are many constraints relevant to the period of transition, which we have discussed in Chapters 7 and 8. As already indicated in this section, our serious and necessary consideration of these constraints has led some reviewers to label us as somewhat ideologically committed to group-based politics. This kind of labelling in South Africa's stressed political debate is inevitable and does not concern us unduly. More important than academic image is the duty to be ruthlessly honest in attempting to understand the constraints and to help reduce the damage these can cause. We will therefore identify four considerations to show that it is only, or even essentially, cultural or racial group identity which makes simple concepts of majority rule inappropriate as a basis for post-apartheid society in the short term.

1. As a society emerging from authoritarian minority domination South Africa will need a period of maturation before political interests can be fully articulated. Current mass-based politics is substantially the politics of liberation, of solidarity, of mobilization against a common enemy, and of needs for redemption; this tradition will take time to be replaced by healthy interest-based political activity in which whites and blacks can find areas of agreement and mutuality. Until then South Africa will remain a politically cleaved society.

   This is why we refer to the need for a period of normalization during which white and black leaders can jointly decide on how to deal with popular expectations.

2. South Africa is a society of dramatic racial inequality, as we have pointed out in the book. There is a danger that class conflict could replace race conflict as a disruptive force. There is a very real danger that new conflicts in a post-apartheid era could continue to frighten off capital investment and retard economic growth. What we require in South Africa is the effective incorporation of both the racial 'classes' in a decision-making system which necessitates innovative strategies and creative compromises to achieve rapid redistribution of resources with the stimulation of growth, and administrative continuity; a principle of healthy constraint stimulating creative strategy for development.

3. A period in which both the old and the new establishments can be incorporated into executive authority will also facilitate nation-building and the emergence of a new South Africa. Both the white minority and the black majority will be reassured and more compliant if their own leaders are seen to be co-operating with one another.

   There are many whites in the Mass Democratic Movement and it is often suggested that they can somehow represent white interests. As white elections have shown for half a century, however, these whites do not represent white political interests or address their sentiments. The whites in the Mass Democratic Movement are an 'intelligentsia', a 'new class' with an internationalist political culture. They are important, indeed vital, components of the new South

Africa but, with rare exceptions, they are almost as foreign to white grassroots politics as the ANC itself.

Another misleading idea is that the protection of the cultural rights of individuals will *ipso facto* secure the compliance of minorities. As the Dutch sociologist Van Amersfoort[3] puts it: 'The characteristic problem for a minority group in the modern democratic state is not so much that it is difficult to secure formal rights, but that the numerical situation restricts the possibility of translating such rights into social influences.'

4. The new climate of reconciliation in government circles should not beguile observers into the belief that the established articulator of white communal interests, the National Party (NP), is about to negotiate itself into the status of a minority opposition party. Minister Gerrit Viljoen, albeit with understatement in deference to the new climate, has been quite explicit. The NP accepts that race classification (the Population Registration Act) will have to go and new methods of incorporating minorities found, but the NP will have . . . 'a very meaningful future role . . . in a future dispensation, probably in . . . some kind of coalition . . .'.[4] The government cannot soften its stand more than this. A nation-wide poll conducted *before* F. W. de Klerk's recent moves showed a 14 per cent increase in support for the white right-wing opposition since the September election, and that well over one-third of white voters felt that the government was on the wrong track or moving 'too fast'.[5] We, and probably the NP as well, are only too aware of the fact that in 1948 the NP was in much the same electoral situation as the Conservative Party (CP) at the moment, and won the election with a mere 14 per cent of white votes. In the former Rhodesia, Garfield Todd's constitutional plan to phase in majority rule was summarily aborted by the unexpected victory of Winston Field's Rhodesian Front in 1961.

For the political change process to attempt to circumvent or eliminate the NPs bottom line may be ideologically attractive, but it would mean prolonging the achievement of a new dispensation for years, during which violent confrontation may once again emerge.

However, communal sentiments and ethnic identities, which most certainly exist and could become more salient in the future (consider the Soviet Union at present), are not necessarily the major immediate concern. The more relevant concern is how to manage the transition to

democracy without alienating the most skilled and experienced echelons in a society that desperately needs them to improve economic
performance. Recently, when Mr Mandela spoke of ANC intentions to
nationalize industry it caused '. . . widespread selling by the same
overseas investors who just last week were clamouring to get into the
local market'.[6] Perceptions of constructive constraints on *both* minority
and majority constituency leadership is perhaps what South Africa
needs most in the present phase. Simultaneously, this will accommodate both majority and minority needs for symbolic reassurance.

Hence we remain more convinced than ever that a 'buffer' period in
which a government of national unity naturally incorporates both
minority and majority leadership into a grand coalition is a vitally
important transitional arrangement, and that the formulation of a final
constitution should not be rushed. Some critics have said that such an
arrangement will simply perpetuate and strengthen racial antipathy and
group polarization. In answer to this we say that if ethnically-based
interests were no more than false political consciousness, a superficial
and insubstantial remnant of history, we would agree with the critics.
We arue, however, that these interests are resilient and deeply rooted
both in structure and sentiment. If a problem of sectional loyalty has
substance and the capacity to do damage, is it best to ignore it, to wish
it away in a welter of (non-racial) idealism? Or, is it better to recognize
these interests and attempt to moderate theirs effects by incorporating
them within a context of trade-offs and allow them to form part of an
arching process of integration?

In this modern world only a fool or a political knave would encourage
sectional nationalism. We suggest that sectional nationalism should be
sublimated and channeled constructively to build a new nation. We are
inclined to insist that the lessons of history and political science reveal
that nations cannot be built on good intentions and unifying rhetoric
alone.

Notes

1   Adam, Heribert, 'The Polish path', in *Southern African review of books,* vol. 3,
    no. 1, October 1989, pp. 3-5.
2   Market and Opinion Research, May 1989.
3   Van Amersfoort, J. M. M., 'Minority as a sociological concept' in *Ethnic and
    racial studies,* vol. 1, no. 2, 1978, pp. 218-34.
4   Patrick Laurence reporting on a press conference by Minister Viljoen, *Satur-
    day Star,* 10 February 1990.
5   Market and Opinion Research, NR 4/89, November 1989.
6   *The Star (Finance)* 13 February 1990.

INDEX